D0849455

THE FALLACY OF FINE-TUNING

VICTOR J. STENGER

THE FALLACY OF FINE-TUNING

WHY THE UNIVERSE IS NOT DESIGNED FOR US

59 John Glenn Drive
Amherst, New York 14228–2119

Published 2011 by Prometheus Books

Inquiries should be addressed to
Prometheus Books
59 John Glenn Drive
Amherst, New York 14228–2119
VOICE: 716–691–0133
FAX: 716–691–0137
WWW.PROMETHEUSBOOKS.COM

15 14 13 12 11 5 4 3 2 1

Library of Congress Cataloging-in-Publication Data

Stenger, Victor J., 1935–
The fallacy of fine-tuning : why the universe is not designed for us / by Victor J. Stenger.
p. cm.
Includes bibliographical references and index.
ISBN 978–1–61614–443–2 (alk. paper)
ISBN 978–1–61614–444–9 (e-book)
1. Causality (Physics) 2. Cosmology. 3. Equilibrium. 4. Religion and science. 5. God—Proof. 6. Atheism. I. Title.

QC6 .4.C3S74 2011
530.01—dc22

2010049901

Printed in the United States of America

To Christopher Hitchens,
whose courage and brilliance I esteem
and whose friendship I treasure.

Contents

Acknowledgments

As with my previous books, I have relied heavily on the advice of others. Once again I have had superb commentary from people with a wide range of expertise, not just in physics but also in astronomy, biology, neuroscience, philosophy, history, and computer science. I want to especially thank physicist Bob Zannelli for his help on all aspects of this work, and for managing the e-mail discussion list avoid-L on which much of the discussion is carried out. Once again, physicist Brent Meeker has been invaluable in his ability to detect errors of science and logic, and for suggesting rewrites for sections that were not adequately clear.

Other members of avoid-L who have made substantial contributions include Richard Branham, Lawrence Crowell, Bill Jefferys, Don McGee, Steven Nunes, Christopher Savage, Sydney Shall, Brian Stilston, and Michael Tutkowski.

I also benefited immensely from the replies to my questions from the following notable physicists and cosmologists: Anthony Aguirre, Sean Carroll, Paul Davies, Roni Harnik, Mario Livio, Martin Rees, and Michael Salem. Of course, any errors in this manuscript are mine alone, and furthermore it should not be assumed that each of the people named agrees with all or any of my conclusions.

I am also very grateful to Paul Kurtz, Jonathan Kurtz, Steven L. Mitchell, and their talented and dedicated staff at Prometheus Books for their continued support of my work. I would like to single out Jade Zora Ballard for her tireless efforts at what must have been a demanding job of copyediting.

Finally, I could never carry out a difficult and time-consuming work such as this without the love and steadfast support of my wife, Phylliss, and our wonderful family.

Preface

In the late nineteenth century in Lithuania, when my paternal grandfather, Anthony Stungaris, was a sixteen-year-old boy, he became gravely ill with diphtheria.[1] His family members laid him out as comfortably as they could in the barn to die. A strong-willed neighbor girl named Benigna, who was two years older, nursed him back to health. They would marry and raise a family. Their first two daughters died in childhood, but they would eventually have six more children. My father, Vytautis, was the third of these, the first son and the last born in Lithuania.

At the time, Lithuania was under the control of the Russian Czar. Anthony worked as a caretaker in a royal forest preserve where no logging was allowed. One winter was so bitterly cold that nearby townspeople were dying, so grandfather allowed them into the forest to gather up dead wood lying on the ground. They did not stop there but cut down some trees as well.

So, as the story goes, my grandfather was sent to Siberia. When he returned a year or so later and could not find work, he decided to emigrate to the United States. He left Benigna and three children behind, promising to send for them once he was settled.

After two years of hearing almost nothing from Anthony, Benigna set out on her own for America, children in tow. Despite not having had a day of schooling (although she spoke several languages, not including English), she managed to get to America and to find her husband in Bayonne, New Jersey. In my house in Colorado I still have the small chest in which they carried all their belongings onboard the ship. My father was only three years old. The year was about 1906.

My grandfather had found a good job with the General Cable Company in Bayonne, and he and my grandmother eventually bought a four-family house on Avenue A, a block from Newark Bay, which in those years was unpolluted. Right down the street from the house was a rocky beach full of broken glass and the decayed Pavonia Yacht Club. When my father and mother married, they moved into one of the cold-water flats on Avenue A. I lived in that house until leaving for graduate school in Los Angeles at age twenty-one, a move my grandmother deeply disapproved of. She had taken my father out of school in the sixth grade so he could work and make money for the family.

My mother was born in Bayonne, her parents emigrating from Hungary. She was one of eleven children and left high school in the tenth grade.

Now, every human being on this planet can tell an interesting story about the events of previous generations that eventually led to their existence. In my case, if my grandmother had not taken the enormous risk of nursing my grandfather, I would not exist. And neither would my daughter, my son, and my four grandchildren. None of us would exist if my grandmother had contracted the disease herself and died. We would not exist if she had not embarked on that complicated journey to America, or if upon getting there she could not find her husband. And what about the many contingencies that led my father and my mother to meet? Carry the story back in time, generation by generation, species by species, until we reach that primordial accident that resulted in the origin of life, and you will realize how lucky each of us is to be here. If we attempt to calculate the a priori probability for all these events happening exactly as they did, we would get an infinitesimally small number.

Many people find it difficult to comprehend how events with very low probabilities can ever happen naturally. The fact of my existence and that of every other human, plant, and animal on Earth is so incredibly unlikely that in many minds it must be the result of some supernatural plan.

In his 1989 tome *The Emperor's New Mind*, mathematician Roger Penrose calculated that the state of the observed universe is one out of ten raised to the power of 10^{123} possible states.[2] We could not even write this number out using every particle in the universe as a zero.

In physics, the state of a multibody system is represented as a point in an abstract multidimensional space called *phase space* where each axis corresponds to one of the degrees of freedom of the system, such as the spatial x-coordinate or y-component of the momentum of a particle. Penrose shows a cartoon of the "Creator" pointing to an absurdly tiny volume in the phase space of possible universes. This is our universe, and while Penrose has remained neutral, the implication drawn by others is that our universe is so wildly unlikely that our very existence provides irrefutable proof of a creator God.

Our current understanding of physics and cosmology allows us to describe the fundamental physical properties of our universe back to as early as a trillionth of a second after it began. At this writing we have two complementary models—the *standard model of particles and forces* and the *concordance model of cosmology* (also referred to as the *standard model of cosmology*)—that successfully describe all the observations made to date of the submicro world (using our highest energy particle accelerators) and of the supermacro world (made with our best space-borne and earthbound telescopes). More advanced instruments, notably the Large Hadron Collider (LHC) now going into operation in Geneva, are expected to extend our description of the universe even further back in time.

The standard models of physics and cosmology depend on a few dozen parameters, such as the masses of the elementary particles, the relative strengths of the various forces, and the cosmological constant. Many of these parameters are very tightly constrained by existing data, which should not be surprising since the models fit the data with good and in some cases great precision. These parameters specify the infinitesimal volume of phase space pointed to by Penrose's "Creator" that defines the state of the universe.

In recent years, people have wondered what the universe would be like if the parameters of the models that describe it were slightly different. Clearly this new set of parameters would specify a different volume in phase space and a different state of the universe. Again, it should come as no surprise that the models describe a universe that would lead to a different set of observations.

One set of observations concerns life on Earth. A number of authors have noted that the universe described when some parameters are slightly changed no longer can support life as we know it. This implies that life, as we know it, depends sensitively on the parameters of our universe, which is unarguable. A more dubious conclusion, which has attracted much theological attention for over two decades now, asserts that the parameters of our universe are "fine-tuned" to produce life as we know it. This is often referred to as the *anthropic principle*, although, as we will see, there are several versions. What could possibly be doing the fine-tuning? Clearly, according to the proponents, it has to be an entity outside the universe, and such an entity is what most people identify as the creator God.

Let us look at a few quotations selected from the vast literature on the subject. Back in 1985, astronomer Edward Robert Harrison wrote:

> Here is the cosmological proof of the existence of God—the design argument of Paley—updated and refurbished. The fine-tuning of the universe provides *prima facie* evidence of deistic design. Take your choice: blind chance that requires multitudes of universes, or design that requires only one.[3]

Geneticist Francis Collins was the head of the Human Genome Project and at this writing directs the United States National Institutes of Health. In his 2006 bestseller, *The Language of God: A Scientist Presents Evidence for Belief,* Collins argues for the following interpretation of the data:

> The precise tuning of all the physical constants and physical laws to make intelligent life possible is not an accident, but reflects the action of the one who created the universe in the first place.[4]

Physician Michael Anthony Corey writes:

> The stupendous degree of fine-tuning that instantly existed between these fundamental parameters following the Big Bang reveals a miraculous level of micro-engineering that is simply inconceivable in the absence of a "supercalculating" Designer.[5]

Astronomer George Greenstein asserts:

> As we survey the evidence, the thought insistently arises that some supernatural agency—or rather Agency—must be involved. Is it possible that suddenly, without intending to, we have stumbled upon scientific proof of the existence of a Supreme Being? Was it God who stepped in and so providentially crafted the cosmos for our benefit?[6]

And theoretical physicist Tony Rothman adds,

> The medieval theologian who gazed at the night sky through the eyes of Aristotle and saw angels moving the spheres in harmony has become the modern cosmologist who gazes at the same sky through the eyes of Einstein and sees the hand of God not in angels but in the constants of nature. . . . When confronted with the order and beauty of the universe and the strange coincidences of nature, it's very tempting to take the leap of faith from science to religion. I am sure many physicists want to. I only wish they would admit it.[7]

Christian philosopher and apologist William Lane Craig has been debating the existence of God and other theological issues for decades. Many transcripts of his debates can be found on his website.[8] I have debated him twice myself, in 2003 at the University of Hawaii in Honolulu and in 2010 at Oregon State University in Corvallis, Oregon. In chapter 6 I will discuss in detail some of Craig's cosmological arguments for a divine creation of the universe.

Craig uses the fine-tuning argument in many of his debates. Here's how he presented it in his 1998 debate with philosopher/biologist Massimo Pigliucci (and in other debates):[9]

> During the last thirty years, scientists have discovered that the existence of intelligent life depends upon a complex and delicate balance of initial conditions given in the Big Bang itself. We now know that life-*prohibiting* universes are vastly more probable than any life-*permitting* universe like ours. How much more probable?
>
> The answer is that the chances that the universe should be life-permitting are so infinitesimal as to be incomprehensible and incalculable. For example, Stephen Hawking has estimated that if the rate of the universe's expansion one second after the Big Bang had been smaller by even one part in a hundred thousand million million, the universe would have recollapsed into a hot fireball.[10] P. C. W. Davies has calculated that the odds against the initial conditions being suitable for

> later star formation (without which planets could not exist) is one followed by a thousand billion billion zeroes, at least.[11] John Barrow and Frank Tipler estimate that a change in the strength of gravity or of the weak force by only one part in 10^{100} would have prevented a life-permitting universe.[12] There are around fifty such quantities and constants present in the Big Bang which must be fine-tuned in this way if the universe is to permit life. And it's not just *each* quantity which must be exquisitely fine-tuned; their *ratios* to one another must be also finely-tuned. So improbability is multiplied by improbability by improbability until our minds are reeling in incomprehensible numbers.

Craig also quotes physicist and prolific author Davies, winner of the 1995 Templeton Prize for Progress in Religion, as saying: "Through my scientific work I have come to believe more and more strongly that the physical universe is put together with an ingenuity so astonishing that I cannot accept it merely as a brute fact."[13]

Craig and many other theist authors frequently cite the 1984 statement by the late Robert Jastrow, former head of NASA's Goddard Institute for Space Studies, as calling this the most powerful evidence for the existence of God ever to come out of science.

> So once again, the view that Christian theists have always held, that there is an intelligent Designer of the universe, seems to make much more sense than the atheistic view that the universe, when it popped into being uncaused out of nothing, just happened to be by chance fine-tuned to an incomprehensible precision for the existence of intelligent life.[14]

As we will see, a lot of science has been done since 1984 and if Jastrow were still alive, I wonder if he would still feel this way. I hope that other physicists and astronomers who may have felt this way a generation ago will take a look at the arguments in this book, many of which have not appeared before.

As a physicist, I cannot go wherever I want to but wherever the data take me. If they take me to God, so be it. I have examined the data closely over many years and have come to the opposite conclusion: the observations of science and our naked senses not only show no evidence for God but also provide evidence beyond a reasonable doubt that a God that plays such an important, everyday role in the universe such as the Judeo-Christian-Islamic God does not exist.[15]

I will devote most of this book to showing why the evidence does not require the existence of a creator of the universe who has designed it specifically for humanity. I will show that the parameters of physics and cosmology are not particularly fine-tuned for life, especially human life. I will present detailed new information not previously published in any book or scientific article that demonstrates why the most commonly cited examples of apparent fine-tuning can be readily explained by the application of well-established physics and cosmology. I will provide references to recent work on the subject by others besides myself that shows it to be very likely that some form of life would have occurred in most universes that could be described by the same physical models as ours, with parameters whose values vary over ranges consistent with those models. And I will show why we can expect to be able to describe any uncreated universe with the same models and laws with at most slight, accidental variations. Plausible natural explanations can be found for those parameters that are the most crucial for life. I will show that the universe looks just like it should if it were not fine-tuned for humanity.

Cosmologists have proposed a very simple solution to the fine-tuning problem. Their current models strongly suggest that ours is not the only universe but part of a *multiverse* containing an unlimited number of individual universes extending an unlimited distance in all directions and for an unlimited time in the past and future. If that's the case, we just happen to live in that universe which is suited for our kind of life. The universe is not fine-tuned to us; we are fine-tuned to our particular universe.

Now, theists and many nonbelieving scientists object to this solution as being "nonscientific" because we have no way of observing a universe outside our own, which we will see is disputable. In fact, a multiverse is more scientific and parsimonious than hypothesizing an unobservable creating spirit and a single universe. I would argue that the multiverse is a legitimate scientific hypothesis, since it agrees with our best knowledge.

In this regard, I should mention that modern *string theory* is used as a possible basis for a multiverse. This view as been promoted by physicist Leonard Susskind, one of the founders of the subject.[16] String theory is the idea that the fundamental "atoms" (uncuttable objects) of the universe are not zero-dimensional particles but one-dimensional vibrating strings.[17] In this theory, the universe has six dimensions beyond the usual four of space and time. The extra dimensions are curled up on such a small scale that we can't detect them, but they are responsible for the "inner" degrees of freedom carried by the atoms, such as spins and electric charge.

Since gravity is included along with the other forces, string theory has been a major candidate for an ultimate unified "Theory of Everything" (TOE). For years, string theorists have been seeking a unique solution to the elegant but enormously complex equations of string theory that would correspond to the universe as we know it, with no adjustable parameters. They have found solutions all right, but no unique one. Instead, the number of vacuum states alone in string theory is 10^{500}. Susskind got the brilliant idea that these 10^{500} solutions is each a different possible universe with different parameters. He pictures these solutions as a landscape with 10^{500} valleys, each valley corresponding to a universe with a particular set of parameters. Assuming a theory, such as inflationary cosmology in which universes are constantly generated by natural quantum processes and fall into a random valley, there is bound to be one universe that has parameters such as ours suitable for life.

Now, I mention this only for completeness. Although, I believe it is adequate to refute fine-tuning, it remains an untested hypothesis. My case against fine-tuning will not rely on speculations beyond well-established physics nor on the existence of multiple universes. I will show that fine-tuning is a fallacy based on our knowledge of this universe alone.

I will also address a number of issues in cosmology and physics that are only indirectly related to fine-tuning but represent major misinterpretations of science by theologians, Christian apologists, and the many layperson authors who are part of the great, richly financed Christian media machine in the United States that promulgates much misinformation about science to the masses. These are necessary to complete my story that not only is there no evidence for God, but also there is strong evidence for his nonexistence beyond a reasonable doubt.

Since the fine-tuning question is basically one of physics, it obviously cannot be understood satisfactorily without some knowledge of physics. I will attempt to elucidate the physics in words that should be understandable to the average reader, although some grasp of scientific terms and methods will be helpful. The boxed equations can be skipped and the general arguments followed from the text. However, to make my case precise I cannot escape some use of mathematics at about the college freshman level or, occasionally, slightly above. Certainly anyone with sufficient knowledge to write authoritatively on fine-tuning should have no trouble following my mathematical arguments.

At the same time, highly trained physicists and cosmologists may not be totally satisfied by my lack of complete mathematical rigor. I will be using what is sometimes called "semi-classical" arguments. For example, at this writing we have no quantum theory of gravity. Instead we have Einstein's general theory of relativity, which still describes all gravitational phenomena after almost a century since it was introduced but is surely not complete. As a separate theory, we have quantum mechanics and its extension to the standard model of particles and forces, which describes everything else prior to data from the LHC. Both of these replace the classical, Newtonian physics that went before. In a semiclassical argument we use classical physics, with some modifications here and there to account for relativistic and quantum effects. I claim this is adequate for my purposes, which only require that I provide a plausible explanation for observations.

I do not have the burden of disproving that God fine-tuned physics and cosmology so that humans formed in his image would evolve. Anyone making such an awesome claim carries the burden of proof. I regard my task as a devil's advocate to simply find a plausible explanation within existing knowledge for the parameters having the values they do. In doing so I will avoid speculating much beyond currently established knowledge, using only the standard models of physics and cosmology and, rarely, slight extrapolations to what can be expected in the next step. For example, I do not use any arguments based on string theory.

I described above Roger Penrose's characterization of fine-tuning in terms of a tiny region of phase space that specified the parameters of our universe. We saw that the region is an infinitesimal volume, one part in ten raised to the power 10^{123}. Those who claim fine-tuning assert that it is demonstrated with that degree of certitude. I will refute this by showing that some form of life would be possible for a wide range of parameters inside a finite volume of phase space.

Soon after finishing the first draft of this manuscript, I learned that philosopher Robin Collins was preparing a book arguing for the existence of God based on fine-tuning. An abridged, but still eighty-two-page-long version was published in 2009.[18] While Collins has a far better understanding of physics than the typical Christian apologist, I think he still exhibits some of the misunderstandings and narrow vision that we will see are common among the proponents of fine-tuning. I will point out a few of these when the subject arises. Also, in a later chapter I will refer to Collins's specific objections to my previously published work.

This also prompts me to make a general observation that is not widely understood by the general public. It is impossible to prove the reality of any god, or anything else for that matter, by deductive logic alone. Any conclusion one makes by deduction is already embedded in the premises made as the first step of that procedure. Thus one has to be very careful about any philosophical argument for God. Unless that argument brings in observational data at some point, the process may be nothing more than the rearranging of words. Having spent a lifetime looking at observational data, you can expect my arguments to be based on science and not philosophical disputation.

On the other hand, it is possible to logically disprove the existence of gods with certain attributes, by showing an inconsistency between those attributes and either the definition of the god or other established facts. For many examples of this, see *The Impossibility of God* by Michael Martin and Ricki Monnier.[19]

1.
Science and God

1.1. NOMA

A widespread belief exists that science has nothing to say about God—one way or another. I must have heard it said a thousand times that "science can neither prove nor disprove the existence of God."

However, in the past two decades believers and nonbelievers alike have convinced themselves in significant numbers, and their opponents in negligible numbers, that the scientific basis for each position is virtually unassailable.

Atheists look at the world around them, with their naked eyes and with the instruments of science, and see no sign of God. Even the most devout theist must admit that the existence of God is not an accepted scientific fact in the same way as, for example, the existence of quarks or black holes. As is the case with God, no one has directly observed these objects. But the indirect empirical evidence is ample for them to be considered to have some relation to reality with a high degree of probability, always with the caveat that future developments could still find a better explanation for this evidence.

Now, the theist will retort that this does not prove that God does not exist. If she is a Christian, she will of course be thinking of the Christian God. But the argument also does not prove that Zeus and Vishnu do not exist, nor Santa Claus and the Tooth Fairy. Still, one can easily imagine scientific experiments to test for the existence of Santa Claus and the Tooth Fairy. Just post lookouts on rooftops around the world on Christmas Eve, and at the bedsides of children who just lost baby teeth.

As I pointed out in my 2007 book, *God: The Failed Hypothesis—How Science Shows That God Does Not Exist*, it is possible to scientifically test the hypothesis of the existence of a god who plays such an active role in the universe as the traditional God of Judaism, Christianity, and Islam.[1] Zeus and Vishnu might be a little tougher to rule out scientifically, but the Judeo-Christian-Islamic God is surprisingly easy to test for by virtue of his assumed participation in every event in the universe, from atomic transitions in distant galaxies to keeping watch that evolution on Earth does not stray from his divine plan.

While the majority of scientists in Western and non-Islamic nations do not believe in God, many prefer to adopt the stance advocated by famed paleontologist Stephen Jay Gould in his 1999 book *Rocks of Ages: Science and Religion in the Fullness of Life*.[2] Gould proposed that we redefine science and religion so that they are "two nonoverlapping magisteria" (NOMA), leaving science to deal with studying nature and religion to deal with morality. Or, to paraphrase Galileo, science tells us the way the heavens go and religion tells us the way to go to heaven.[3]

Gould's NOMA especially appeals to believing scientists. The many I have known in fifty years of academic life place their religion and science in separate compartments that never interact with one another. Most nonbelieving scientists go along with NOMA as well, since they would prefer, as a social and political strategy, to avoid getting into battles over religion. However, philosophers, theologians, and many atheist scientists have not found NOMA practical. Notice I referred to Gould's proposal as a "redefinition." Gould, an avowed atheist, now deceased, was good-intentionally trying to carve out distinct, "nonoverlapping" areas for both religion and science. He strived to eliminate conflicts, which had increased in recent times, by basically redefining religion as moral philosophy. However, existing religions, while claiming to tell us how to go to heaven, also try to tell us how the heavens go. Moreover, science is not proscribed from observing human behavior and providing observational data on matters of morality.

NOMA simply does not accurately describe either the history or the current status of the relationship between religion and science. Neither is likely to agree to any limitations on its zone of activity. So they overlap and are going to continue to do so, and they will continue to battle over their common ground where differences are, in many cases, irreconcilable.

1.2. NATURAL THEOLOGY

During the Enlightenment in the eighteenth century, when the rise of science increasingly influenced thinking, religion was deeply questioned in Europe and America. For the first time perhaps in history, it became possible to be openly atheistic or at least critical of established religion. But it was not a one-way street. Western Christian theology, which by then already had a proud history of logical thinking on the problem of God, found a place for science in what was called *natural theology*.

Natural theology provided several excellent scientific arguments for the existence of God that, when first introduced, were irrefutable with existing scientific knowledge. They only became refutable with further scientific developments.[4]

The premier figure in natural theology was William Paley (d. 1805), archdeacon of Carlisle, whose 1802 book, *Natural Theology; or, Evidences of the Existence and Attributes of the Deity*, was the first serious attempt to use scientific arguments to prove that the world was designed and sustained by God.[5] While design arguments for God had been proposed since antiquity, and argued against—notably by the great Scottish philosopher David Hume (d. 1776) in his *Dialogues concerning Natural Religion*[6] and *An Enquiry concerning Human Understanding*[7]—Paley went beyond the typical theological emphasis on logical deduction to a direct appeal to empirical observation and its interpretation.

This was important because every one of the endless series of "proofs" of the existence of God that has been proposed, from antiquity to the present day, is automatically a failure because, as I have mentioned, a logical deduction tells you nothing that is not already embedded in its premises. All logic can do for you is test the self-consistency of those premises. There is only one reliable way that humans have discovered so far to obtain knowledge they do not already possess—observation. And science is the methodical collecting of observations and the building and testing of models to describe those observations.

Paley's main argument was based on the "watchmaker analogy," which had been used by others in the past to illustrate divine order in the world. In his opening paragraph he talks about crossing a heath and pitching his foot against a stone. He sees no problem thinking that the stone had not lain there forever. On the other hand, if he had found a *watch* upon the ground and saw that its several parts were put together for a purpose, the inference would be inevitable that the watch must have had a maker, an artificer who had formed it for a purpose. He then proceeds to make an analogy between the watch and living creatures, with their eyes and limbs so intricately designed as to defy any imaginable possibility that they could have come about by any natural process.

1.3. DARWINISM

This argument convinced almost everyone, even the young Charles Darwin (d. 1882) when he was a student, coincidentally occupying the same rooms at Cambridge as Paley had a generation earlier. But Darwin would eventually change many minds besides his own. During his voyage around the world on HMS *Beagle* from December 27, 1831, to October 2, 1836, Darwin accumulated a wealth of data that he analyzed meticulously during the next twenty-three years. In 1859, he published *On the Origin of Species: By Means of Natural Selection, or the Preservation of Favoured Races in the Struggle for Life*, which demonstrated how, over great lengths of time, complex life-forms evolve by a combination of random mutations and natural selection.[8] Living organisms not only develop without the need for the intervention of an intelligent designer but also provide ample evidence for the lack of such divine action.[9]

While it should not be forgotten that Alfred Russel Wallace announced his independent discovery of natural selection simultaneous with Darwin (by mutual agreement), and others had also toyed with the notion, Darwin deserves the lion's share of the credit by virtue of providing the great bulk of supporting evidence and his brilliant insights in interpreting that evidence.

I need not relate the familiar history of the 150-year battle between science and religion over the theory of evolution, especially the attempts by Christians to have public schools in the United States teach the biblical creation myth as a legitimate "scientific" alternative.[10] While heroic attempts have been made by theists and atheists alike to show that evolution need not conflict with traditional beliefs, the fact remains that the majority of believers in the United States refuse to accept a scientific theory that is as well established as the theory of gravity because of its gross conflict with the biblical account of the creation of life.

A series of Gallup polls of Americans from 1982 to 2008 asked respondents to choose from three options: (1) Humans developed over millions of years, God-guided, (2) Humans developed over millions of years, God had no part, (3) God created humans as is within ten thousand years. The results were fairly consistent over the years, the 2008 results giving 36 percent for God-guided but over millions of years, 14 percent for the long period with God having no part, and 44 percent with creation as is within the last ten thousand years.[11]

Another recent poll was conducted by Vision Critical, a UK organization, on the question of whether human beings evolved from less-advanced forms over millions of years or whether they were created by God in their present form within the last ten thousand years. The result for Great Britain was that 68 percent supported evolution, 16 percent supported creation, and 15 percent were unsure. The result for Canada was 61 percent for evolution, 24 percent for creation, and 15 percent unsure. The result for the United States was 35 percent for evolution, 47 percent for creation, and 18 percent unsure.[12] The difference between the United States and the two nations closest to it in culture is striking.

Only the Gallup poll considered the question of God guidance. While it is true that there were people before Darwin, including his own grandfather, who had speculated about evolution, today the term is understood to include the Darwin-Wallace mechanism of random mutations and natural selection. There is no crying in baseball, and there is no guidance, God or otherwise, in Darwinian evolution. Only the 14 percent of Americans who accept that God had no part in the process can be said to believe in the theory of evolution as the vast majority of biologists and other scientists understand it today. God-guided development is possible, but it is unnecessary and just another form of intelligent design.

Just because the Catholic Church and moderate Protestant congregations *say* they have no problem with evolution, that doesn't mean they don't. A statement by Pope John Paul II in 1996 seemed to support biological evolution. However, he made it clear that in his opinion it was still one of several hypotheses still under dispute. That opinion sharply disagrees with that of the vast majority of biologists. Furthermore, the pope unambiguously excluded the evolution of mind, saying that "the spiritual soul is immediately created by God" and that theories of evolution that consider mind as emerging from living matter "are incompatible with the truth about man."[13] No doubt the pope has never considered the possibility that the evolution of the human species was not controlled by God.

In the theory of evolution accepted by an almost unanimous consensus of scientists, humans with fully material bodies evolved by accident and natural selection only, with no further mechanisms or agents involved, and simply were not designed by God or natural law.[14] The evolution of mind is currently more contentious, but the evidence piles up daily that mind is also purely the product of the same natural processes with no need to introduce anything beyond matter. This conclusion is unacceptable to anyone who has been raised to think he was made in the image of God with an immortal, immaterial soul that is responsible for our conscious thinking.

It is important to recognize that when evolution by natural selection was first proposed in 1859, it was not in agreement with all scientific knowledge and was potentially falsifiable. According to calculations by the great physicist William Thomson, Lord Kelvin, the sun did not have enough stored energy to last the millions of years needed for biological evolution. It was not until the early twentieth century that nuclear energy was discovered and was shown to be the highly efficient source of energy of the sun and other stars that allows them to shine for billions of years, thus providing ample time for life to evolve.

1.4. INTELLIGENT DESIGN

In recent years we have seen Paley's argument exhumed with an attempt to place it on a sounder scientific basis, or, at least, to make it seem so. In 1996, biochemist Michael Behe published *Darwin's Black Box: The Biochemical Challenge to Evolution*, which claimed that some biological structures are "irreducibly complex."[15] That is, living systems possess parts that could not have evolved from simpler forms since they had no function outside of the system of which they were part. His examples included bacterial flagella and blood clotting. Evolutionary biologists, of whom Behe is not one, easily demonstrated the flaw in this argument. Parts of biological structures often evolve with one function and then change function when joining up with another system. This was well known before Behe wrote his book, and many examples have since been described.

In 1999, theologian William Dembski published a book called *Intelligent Design: The Bridge between Science & Theology*, claiming that he could mathematically demonstrate that living systems contained more information than could be generated by natural means alone.[16] While he had a number of other arguments based on information theory, they all boiled down to what he called the "law of conservation of information." This law, Dembski asserted, required that the amount of information output by a physical system could never exceed the amount of information input. Thus it followed that the large amounts of information contained in living systems must have had an external input of information provided by an intelligent designer outside nature, who shall remain nameless.[17]

Hundreds of papers and dozens of books have refuted Dembski (as well as Behe), and I need not refer to them all. I will just mention his misuse of the concept of information.[18] Dembski used the definition of information provided by the father of communication theory, Claude Shannon, in 1948.[19] Shannon defined the information transferred in a communication process to be equal, within a constant, to the decrease in the entropy of the system. Here he used the conventional definition of entropy in statistical mechanics that was provided by Ludwig Boltzmann in the late nineteenth century.

Now, it is well known that entropy is not a conserved quantity such as energy. The second law of thermodynamics allows for the entropy of a closed system to increase with time. It follows that information is not a conserved quantity and Dembski's law of conservation of information is provably wrong. On the empirical side, many examples can be given of physical systems creating information. A spinning compass needle provides no information on direction. When it slows to a stop, it "creates" the information of the direction North.

The particular form of intelligent design proposed by Behe and Dembski received a deadly blow in December 2005, when a federal court in Dover, Pennsylvania, ruled that it was motivated by religion and thus would violate the Establishment Clause of the US Constitution if taught as science in public schools.[20] There can be no doubt that intelligent design claims are motivated by religion. However, in his ruling the judge went further than necessary by declaring that intelligent design is not science. It is my professional opinion and that of several philosophers that intelligent design is in fact science, just wrong science. That should be sufficient to keep it out of classrooms along with phlogiston and the theory that Earth is flat, except as historical references.

2.
The Anthropic Principles

2.1. FINE-TUNING

For years now theists have thought they have the final, killer scientific argument for the existence of God. They have claimed that the physical parameters of the universe are delicately balanced—"fine-tuned"—so that any infinitesimal changes would make life as we know it impossible. Even atheist physicists find this so-called "anthropic principle" difficult to explain naturally, and many think they need to invoke multiple universes to do so.

2.2. HISTORY

Let us review the history of the notions of fine-tuning and the anthropic principles. In 1919, physicist Hermann Weyl expressed his puzzlement that the ratio of the electromagnetic force to the gravitational force between an electron and a proton is such a huge number, $N_1 = 10^{39}$.[1] Weyl wondered why this should be the case, expressing his intuition that "pure" numbers occurring in the description of physical properties, such as π, which do not depend on any system of units, should most naturally occur within a few orders of magnitude of unity. Unity, or zero, you can expect "naturally." But why 10^{39}? Why not 10^{57} or 10^{-123}? Some principle must select out 10^{39}. This is called *the large number puzzle*.

In 1923 Arthur Eddington commented: "It is difficult to account for the occurrence of a pure number (of order greatly different from unity) in the scheme of things; but this difficulty would be removed if we could connect it to the number of particles in the world—a number presumably decided by accident."[2] He estimated that number, now called the "Eddington number," to be $N = 10^{79}$. Well, N is not too far from the square of N_1. This coincidence was no doubt accidental, since we now know that it corresponds just to the number of atoms in the visible universe, which contains a billion times as many photons and neutrinos and many billions of times more beyond our horizon.[3]

These musings may bring to mind the measurements made on the Great Pyramid of Egypt in 1864 by the Astronomer Royal for Scotland, Charles Piazzi Smyth. He found accurate estimates of π and the distance from Earth to the sun, and other strange "coincidences" buried in his measurements.[4] However, further analysis revealed that these were simply the result of Smyth's selective toying with the numbers.[5] Still, even today some people believe that the pyramids hold secrets about the universe. Ideas like this never seem to die, no matter how deep in the Egyptian sand they may be buried.

Look around at enough numbers and you are bound to find some that appear connected. Most physicists, therefore, did not seriously regard the large number puzzle until one of their most brilliant members, Paul Dirac, took an interest. Few physicists ignored anything Dirac had to say.

Dirac pointed out that N_1 is the same order of magnitude as another pure number, N_2, that gives the ratio of a typical stellar lifetime to the time for light to traverse the radius of a proton. That is, he found two seemingly unconnected large numbers to be of the same order of magnitude.[6] If one number being large is unlikely, how much more unlikely is another to come along with about the same value?

In 1961 physicist Robert Dicke pointed out that N_2 is necessarily large in order that the lifetime of typical stars be sufficient to generate heavy chemical elements such as carbon. Furthermore, he showed that N_1 must be of the same order as N_2 in any universe with heavy elements.[7] This was the first of the anthropic coincidences: if N_1 did not approximately equal N_2, life as we know it would not exist.

The heavy elements did not get fabricated straightforwardly. According to the big bang theory, only hydrogen, deuterium (the isotope of hydrogen containing one proton and one neutron in its nucleus), helium, and lithium were formed in the early universe. Carbon, nitrogen, oxygen, iron, and the other elements of the chemical periodic table were not produced until billions of years later. These billions of years were needed for stars to form and in their death throes, after burning all their hydrogen, to assemble these heavier elements out of neutrons and protons. When the more massive stars expended their hydrogen fuel, they exploded as supernovae, spraying the manufactured elements into space. Once in space, these elements cooled, mixed with the interstellar medium, and eventually formed newer stars accompanied in many instances by planets.

Billions of additional years were needed for our home star, the sun, to provide a stable output of energy so that at least one of its planets could develop highly complex life. But if the gravitational attraction between protons in stars had not been many orders of magnitude weaker than the electric repulsion, as represented by the very large value of N_1, stars would have collapsed and burned out long before nuclear processes could build up the periodic table from the original hydrogen and deuterium. The formation of chemical complexity is possible only in a universe of great age in terms of nuclear reaction times—or at least in a universe with other parameters close to the values they have in this one.

2.3. HOYLE'S PREDICTION

The next important step in the history of fine-tuning occurred in 1952 when astronomer Fred Hoyle used anthropic arguments to predict that an excited carbon nucleus, ${}_6C^{12}$, has an energy level at around 7.7 MeV. Hoyle had looked closely at the nuclear processes involved and found that they appeared to be inadequate.

The basic mechanism for the manufacture of carbon is the fusion of three helium nuclei into a single carbon nucleus:

$$3{}_2He^4 \rightarrow {}_6C^{12}$$

The subscript on the chemical symbol tells us how many protons are in the nucleus: 2 for ${}_2He^4$, 6 for ${}_6C^{12}$. This is called the *atomic number* and is the position number in the periodic table. Although it is redundant with the chemical symbol, I have included it for pedagogical purposes. The superscripts give the total number of *nucleons*, that is, protons plus neutrons in each nucleus. This is called the *nucleon number* or *mass number* and relates to the *atomic weight* in chemistry. The total number of nucleons is *conserved*, that is, remains constant, in a nuclear reaction. So does the total atomic number, since it measures nuclear electrical charge and this is also conserved. Chemical elements with the same atomic number but different atomic weights are called *isotopes*.

The probability of three bodies coming together simultaneously is very low, and some catalytic process in which only two bodies interact at a time must be assisting. An intermediate process in which two helium nuclei first fuse into a beryllium nucleus, which then interacts with the third helium nucleus to give the desired carbon nucleus, had earlier been suggested by astrophysicist Edwin E. Salpeter:[8]

$$2\,{}_{2}He^{4} \rightarrow {}_{4}Be^{8}$$
$${}_{2}He^{4} + {}_{4}Be^{8} \rightarrow {}_{6}C^{12}$$

Hoyle showed that this still was not sufficient unless the carbon nucleus had an excited state at 7.7 MeV to provide for a high reaction probability. A laboratory experiment was undertaken, and, sure enough, a previously unknown excited state of carbon was found at 7.65 MeV.[9]

Nothing can gain you more respect in science than the successful prediction of a new phenomenon. Hoyle's prediction provided scientific legitimacy for anthropic reasoning. But he also gave believers (he claimed to be an atheist himself) something to crow about, remarking,

> A commonsense interpretation of the facts suggests that a superintellect has monkeyed with physics, as well as with chemistry and biology, and that there are no blind forces worth speaking about in nature. The numbers one calculates from the facts seem to me so overwhelming as to put this conclusion almost beyond question.[10]

2.4. THE ANTHROPIC PRINCIPLES

In 1974, physicist Brandon Carter introduced the term *anthropic principle* to describe the anthropic coincidences.[11] In the *weak* form of the principle, the location in the universe in which we live must be compatible with the fact that we are here to observe it. In the *strong* form, at least at some stage, the universe itself must have been compatible with the existence of observers.

In its weak form, the anthropic principle simply points out the obvious fact that if the laws and parameters of nature were not suitable for life, we would not be here to talk about them. As a simple example, physical constants and the laws that contain them determined that the atmosphere of Earth would be transparent to wavelengths of light from about 350 nanometers to 700 nanometers, where a nanometer is a billionth of a meter. The human eye is sensitive to the same region. The theistic interpretation is that God designed our eyes so we could see in front of us. The natural explanation is that our eyes evolved to be sensitive in that region.

One possible natural explanation for the anthropic coincidences is that multiple universes exist with different physical constants and laws and our life-form evolved in the one suitable for us. Theists vehemently object that we have no evidence for multiple universes and, furthermore, we are violating Occam's razor by introducing multiple entities "beyond necessity." Even some atheistic physicists criticize the idea as "nonscientific." However, cosmologist and evangelical Christian Don Page, who was Stephen Hawking's student, finds the idea of multiple universes congenial to theism. He suggests, "God might prefer a multiverse as the most elegant way to create life and the other purposes He has for His Creation."[12]

Modern cosmological theories do indicate that ours is just one of an unlimited number of universes, and theists can give no reason for ruling them out. In this book I do not spend a lot of time with philosophical or "commonsense" reasoning. Rather I look directly at the various parameters that have been proposed as being fine-tuned and see if plausible explanations can be found within existing knowledge.

None of the books and articles I have seen that promote the anthropic coincidences as scientific evidence for God, including those written by scientists, present precise scientific arguments. Although many journal articles have been published on special topics, the only place one can find a detailed discussion of the physics of fine-tuning, with equations, is in the 1986 tome by physicists John Barrow and Frank Tipler, *The Anthropic Cosmological Principle.* This book provides an exhaustive historical, philosophical, and scientific survey of the notion that the universe is fine-tuned for humanity, without taking a stand one way or the other.[13]

Barrow and Tipler defined three different forms of the anthropic principle, the first of which is as follows:

Weak Anthropic Principle (WAP):

> The observed values of all physical and cosmological quantities are not equally probable but they take on values restricted by the requirement that there exist sites where carbon-based life can evolve and by the requirement that the Universe be old enough for it to have already done so.[14]

This is essentially the same as Carter's definition. As we saw above, all the WAP seems to say is that if the universe were not the way it is, we would not be here talking about it. If the mass of the electron were different, people would look different. However, it does not tell us why the constants have the values they do rather than some other value that would make life impossible.

Barrow and Tipler formulated the strong anthropic principle as follows, which differs from Carter's definition:

Strong Anthropic Principle (SAP):

> The universe must have those properties that allow life to develop within it at some stage in its history.[15]
>
> Barrow and Tipler noted that SAP can have three interpretations:

1. *There exists one possible universe "designed" with the goal of generating and sustaining "observers."*

2. *Observers are necessary to bring the universe into being.*
3. *An ensemble of other different universes is necessary for the existence of our universe.*

The authors identify a third version of the anthropic principle:

> *Final Anthropic Principle* (FAP):
>
> Intelligent, information-processing must come into existence in the universe, and, once it comes into existence, it will never die out.

The late polymath Martin Gardner referred to this as the *Completely Ridiculous Anthropic Principle* (CRAP).[16]

2.5. FINE-TUNING TODAY

Since Barrow and Tipler, many books and articles have appeared that have made much of the anthropic coincidences. As we have seen from the several quotations I have already presented, believers have little doubt that here is indisputable evidence for a creator. William Lane Craig's mind "reels" at the incomprehensibly small probabilities for any natural explanation. He says there are fifty fine-tuned parameters, but he does not list them. A list of thirty-four parameters that are claimed to be fine-tuned has been assembled by microbiologist Rich Deem on his *God and Science* website.[17] Deem also lists, without details, estimates of the precision at which each parameter had to be tuned to produce our kind of life. These are the numbers that make Craig's mind reel.

Deem's main reference is physicist and Christian apologist Hugh Ross and his popular book *The Creator and the Cosmos*, first published in 1993.[18] Ross is the founder of Reasons to

Believe, which describes itself as an "international and interdenominational science-faith think tank *providing powerful new reasons from science to believe in Jesus Christ*."[19] A list of twenty-six claimed "design evidences" can be found in the book.[20] Ross has further developed his arguments in a chapter called "Big Bang Model Refined by Fire" in the anthology *Mere Creation: Science, Faith & Intelligent Design*.[21]

Without giving any more quotations, I will just list a few additional books by believers in my private library that call on the anthropic coincidences to make their case for God:

- *Atheism Is False*, by David Reuben Stone[22]
- *A Case against Accident and Self-Organization*, by Dean L. Overman[23]
- *Life after Death*, by Dinesh D'Souza[24]
- *The Language of God*, by Francis Collins[25]

These and many more can be found in university and public libraries, and I have studied a sufficient number of these efforts to have a good grasp of the claims being made.

In addition, a number of reputable, nontheistic scientists have written popular books on fine-tuning. They try to keep an open mind on the theological implications but still express puzzlement alongside the enthusiasm typical of science writers anxious to exhibit the wonders of the universe to their readers. Again, in my private library I find:

- *The Constants of Nature*, by John Barrow[26]
- *Just Six Numbers*, by Martin Rees[27]
- *The Goldilocks Enigma*, by Paul Davies[28]

Needless to say, theologians and philosophers have not ignored fine-tuning. Perhaps the best-known theologian arguing for the existence of God is Richard Swinburne. His article, "Argument from the Fine-Tuning of the Universe,"[29] and a review, "The Anthropic Principle Today,"[30] by philosopher John Leslie can be found in a very useful set of essays *Modern Cosmology & Philosophy*, edited by Leslie.[31]

In this book I will look in detail at all of the most significant claims of fine-tuning. But first I need to establish some basic facts about physics and cosmology that are much misunderstood, or at least misrepresented, especially in the Christian apologetic community.

3. The Four Dimensions

3.1. MODELS

The process of doing physics is no different than that of any other natural science. We make observations, almost exclusively quantitative measurements, and describe these observations with mathematical *models*. We attempt to use those models to correlate and predict other observations. When a model's predictions are risky; that is, when they have a good chance of turning out wrong, then the model can be fairly tested by checking those predictions. If the test fails, the model is falsified and we forget about it, or, more frequently, we modify it and try again. Models that cannot be falsified, that explain everything, explain nothing.

At some point, after a model has passed many risky tests, it may be granted the exalted status of "theory." The two models we will discuss the most in this book, the *standard model of particles and forces* and the *standard (concordance) model of cosmology*, have legitimately achieved theory status by their success at making many risky predictions. Nevertheless, they are still referred to as "models." This may be the unfortunate consequence of creationists' attempt to demean Darwinian evolution in the public's mind by referring to it as "just a theory." Or perhaps it is just habit.

Scientists and philosophers of science have never been able to agree on a demarcation criterion that precisely distinguishes science from nonscience. The falsification criterion proposed by Karl Popper and Rudolf Carnap is still commonly referred to but has proved unsatisfactory, at least to most philosophers.[1] While not everything that is falsifiable, such as astrology, is science, my own experience in physics research has been that the failure of a logically consistent model to agree with observations is the only reason that we have to discard the model. Without falsification science would be an anarchy of logically consistent but still useless models that simply suit someone's fancy. But falsification does not solve the demarcation problem.

As is often pointed out, since we can never know whether a model might be falsified in the future, its current agreement with the data is never sufficient to verify the model with 100 percent certainty. However, the admitted fact that science is always tentative is often overemphasized to the point where laypeople wonder if science can say anything certain. When a model has passed many risky tests, tests that could have falsified it, we can begin to have confidence that it is telling us something about the real world with certainty approaching 100 percent.

It's not likely that we will someday discover that Earth is flat after all, or that it really is only six thousand years old. In cases like these we can say that science has "proved" something "beyond a reasonable doubt." This still leaves open a tiny possibility for change but assures us we can take these as scientific facts for all practical purposes. If we were going to take too literally the statement that science is tentative, we would never set foot in an airplane nor allow ourselves to be put to sleep for surgery.

A good example of a model that we can accept with some confidence is inflationary cosmology, which will be discussed in chapter 5. This model has passed a number of tests that could have falsified it. In the meantime, it helps explain much of what is observed in cosmology, including, as we will see, several observations that theists insist are examples of fine-tuning. However, being relatively recent and still having some theoretical problems, inflationary cosmology remains subject to further testing and consideration of alternatives.

3.2. OBSERVATIONS

Even if we cannot precisely distinguish science from nonscience, we can establish several facts about the scientific process. Most important of all, science deals with observations. If you are talking about science, you are talking about data. If you are not talking about data, you are not talking about science.

Now, it is true that string theory and other theoretical attempts to unite general relativity and quantum mechanics have yet to confront empirical tests. However, they must ultimately do so or cease to be considered scientific endeavors.

Any process must start with some definitions. The physical science process starts with *operational definitions* of the quantities that will be part of the models that will be built to describe observations. The operational definition of a quantity prescribes the precise procedure that one goes about in making a measurement of that quantity. For example, before the establishment of the atomic theory of matter, the absolute temperature of a body was defined as what is measured with a *constant volume gas thermometer*. This made it possible for workers in one laboratory to present their results in terms of a quantity that could be unambiguously tested in any other laboratory with the same class of instrument. Other types of thermometers might be used, but these would have to be calibrated against the constant volume gas thermometer. And that was temperature, by definition. Temperature was not some Platonic form out there in a metaphysical world beyond the laboratory. It was what you measured on a constant volume gas thermometer.

3.3. SPACE, TIME, AND REALITY

Most people, including most physicists, believe that the models and "laws" of physics directly describe reality. That is, the objects in these models actually exist outside the paper they are written on and the concepts they contain refer directly to true aspects of the world. Consider space and time, the two concepts that form the starting point of all physical models.[2] What could be more real than space and time?

Originally, space-time models assumed three dimensions of space obeying *Euclidean* geometry, with time an independent measure. In 1905, Einstein introduced a model called *special relativity* that united space and time in a single four-dimensional manifold we call space-time. In 1916, he introduced *general relativity* that utilized a more general *non-Euclidean* geometry that had been developed by mathematicians in the previous century. The surface of a sphere was an example of non-Euclidean geometry, where the parallel lines of longitude meet at the poles. Thus Einstein formulated the idea that in the vicinity of a heavy mass space is "curved" and that bodies naturally followed "geodesics" analogous to the great circles on a sphere. He thereby accounted for gravity without introducing a gravitational force. Like an airliner taking the shortest available distance, a great circle from point to point on Earth's surface, the geodesic path is the path of "least action."

Today the most fashionable model being worked on in physics is called *M-theory* (a generalized form of *string theory*) in which space-time has ten dimensions of space and one of time. Seven of the ten spatial dimensions are curled up on such a tiny scale as to be invisible to even our most powerful instruments, but they have degrees of freedom that give particles their "internal" properties, such as spin and electric charge.

In his popular book on fine-tuning, *Just Six Numbers*, the Astronomer Royal of the United Kingdom, Martin Rees, proposes that the dimensionality of the universe is one of six parameters that appear particularly adjusted to enable life.[3] He presents arguments known from the time of Newton that three spatial dimensions are special. In particular, in three dimensions gravity obeys an inverse square law, without which stable planetary orbits would not be possible.

Clearly Rees regards the dimensionality of space as a property of objective reality. But is it? I think not. Since the space-time model is a human invention, so must be the dimensionality of space-time. We choose it to be three because it fits the data. In the M-theory we choose it to be eleven. We use whatever works, but that does not mean reality is exactly that way in one-to-one correspondence.

The same thing can be said about the geometry of space. That's our invention, too. It happens to be non-Euclidean in Einstein's beautiful model, but perhaps someday a simpler, Euclidean model will be found with gravity treated in a different way.

When physicists talk about space and time, as in general relativity and M-theory, they implicitly assume that they are describing reality. At least, they talk as if they are. Time is "real." Space is "real" and it "really" has three, ten or some other number of dimensions still to be determined by experiment. By "real" here I am referring to ontology, not mathematics.

But how do physicists know that space, time, or any other of their invented quantities are real? All they can know is that they have invented some models with which they are trying to describe observations. If they agree with observation, then the models no doubt have something to do with reality. However, do physicists have any basis for assuming that the models and "laws" of physics, which are human inventions, describe reality precisely as it is? Why couldn't reality be something totally different that gives the same observations?

This was certainly Plato's view in his metaphor of the cave. Prisoners are chained inside a cave so they can see only the wall. Their only knowledge of what is going on outside their view is obtained from the shadows on the wall. Are our observations just shadows of what is truly real? We can never know.

Do we have any basis for assuming that the models and "laws" of physics, which are human inventions, describe reality as it is? The success of quantum mechanics has made that question a lot more profound. Is an object "really" a particle or a wave? Does a particle "really" have a momentum after its position has been measured? Is this particle's wave function a "real" field that collapses instantaneously far out to the edge of the universe and back all the way in the past when the position is measured here and now? If the wave function is not real, then what about the electric and magnetic fields that are associated in quantum mechanics with the wave function of a photon? Does any physicist consider them *not* real? No, but they can't prove it.

A common refrain among theoretical physicists is that the fields of quantum field theory are the "real" entities while the particles they represent are images like the shadows in Plato's cave. As one who did experimental particle physics for forty years before retiring in 2000, I say, "Wait a minute!" No one has ever measured a quantum field, or even a classical electric, magnetic, or gravitational field. No one has ever measured a *wavicle*, the term used to describe the so-called wavelike properties of a particle. You always measure localized particles. The interference patterns you observe in sending light through slits are not seen in the measurements of individual photons, just in the statistical distributions of an ensemble of many photons. To me, it is the particle that comes closest to reality. But then, I cannot prove it is real either. I will expand on this point in chapter 15.

We have to think about what it is we actually do when we make measurements and fit them to models. We do not use the models to build up some metaphysical system. We use the models to predict the outcomes of other measurements and to aid us in putting these to practical use. Consider the example of transmitting an electromagnetic signal from one point to another in empty space. For simplicity, let us try to send a signal of a single frequency. We set up an oscillating electric current in an antenna and tune a receiver attached to a distant antenna to the frequency of that oscillation. We then use Maxwell's equations to calculate the form of the electromagnetic wave that will propagate through space at the speed of light from the first antenna to the second. But we never see that wave. We just measure an oscillating current in the receiver. In other words, it does not matter whether or not the electromagnetic field is "real." It is part of a model that we use to describe the data but it never enters into the measurement process at either end of the signal.

Now, none of this should be interpreted as meaning that physics is not to be taken seriously. When I say physical models are human inventions, I mean the same as if I were saying that the camera is a human invention. Like the camera, the models of physics very usefully describe our observations. When they do not, the model or the camera is discarded. I am simply repeating what many philosophers have pointed out over the centuries, that our observations are not pure but are operated on by our cognitive system composed of our senses and the brain that analyzes the data from those senses. Those models need not correspond precisely, or even roughly, to whatever reality is out there—although they probably do at least for large objects. The moon is probably real. But the gravitational field does not have to be.

Now, this may seem like pedantic philosophizing, but it is important when we start talking about God fine-tuning the parameters of our models. Why should quantities that are simply human artifacts used in describing nature have to have external forces setting their values?

Still, I realize that I open myself up to some tough questions by taking this point of view. If the models and parameters are just human inventions, why should they have anything to do with objective reality? Well, they are not arbitrary since they have to agree with observations, not just roughly but with quantitative accuracy. Furthermore, I have already admitted that the moon is probably real. Where do I draw the line? Let's say macroscopic bodies that we see with unaided eyes are real. Does that mean that bacteria we can see only with a microscope are imaginary? No biologist would let me get away with that.

It is not until you get to the "submicroscopic" quantum level that the reality issue comes up. There, our models include things such as "virtual particles" with imaginary mass and wave functions that propagate instantaneously throughout the universe. Later, after we have developed physics ideas further, I will delve a little into speculative metaphysics just to show that a plausible and consistent, if unprovable, picture exists for the reality behind observations.

3.4. PARAMETERS

The properties of the universe that are supposed to be fine-tuned for life arise from how matter is described in our models from the tiniest distances at subnuclear level to the greatest distances in the cosmos. The quantities that theists refer to when they talk about fine-tuning are usually either "constants" in the models of physics, such as the masses of particles and the relative strengths of forces, or physical properties used in the models of cosmology, such as the average mass density of the universe or the cosmological constant. It is also often suggested that the models of physics themselves are fine-tuned—that they might have been different and so needed the special attention of a creator to come out "just right." But if they are human inventions, then they needed the special attention of a human to come out "just right."

Now, since we are allowing the quantities of physics, such as the masses of particles, to take on different values, it is confusing to refer to them as "constants" and so I will call them "parameters." Indeed, unbeknownst to most of the non-physicist authors who write about fine-tuning, some of the quantities, such as the relative strengths of forces, are not even constant in our universe but depend on the energies of the particles interacting with one another. I will discuss this in detail in chapter 10.

3.5. DEFINITIONS

Space and time are the two concepts that form the starting point of virtually all physical models. Over the centuries, the standard units of distance and time have been steadily refined to be as convenient and stable as possible. In 1960, the meter was defined as 1,650,763.73 wavelengths in a vacuum of the electromagnetic radiation that results from the transition between the 2p10 and 5d5 energy levels of the Krypton-86 atom. In 1967, the second was defined as 9,192,631,770 periods of the radiation corresponding to the transition between the two hyperfine levels of the ground state of the Cesium-133 atom.

The quantity conventionally labeled *c* in physics is called the "speed of light in a vacuum." In 1905, Einstein proposed that *c* is a universal constant, and this became one of the axioms of his *theory (model) of special relativity*. By 1983, special relativity had proved so successful that the universality of *c* was accepted as a fact by the scientific community and built into the very definitions of space and time.

In the Standard International (SI) system of units, the distance between two points in space is measured in *meters*. Until 1983, the meter was defined independently of the second. In that year, by international agreement, it was mandated that the meter would be defined as the distance between two points when the time it takes light to go between the points in a vacuum is 1/299,792,458 second. That is, the speed of light in a vacuum is $c = 299{,}792{,}458$ meters per second *by definition*.

In general, then, time is what you measure on a clock and

distance is what you measure on the *same* clock for light to go between two points. The significance of this agreement has never been fully understood, either by laypeople or by many physicists and philosophers. The quantity c cannot be fine-tuned. It is fixed by definition.

But how can this be? How can the speed of some physical process be simply declared by fiat? Light travels so many meters per second. Isn't that a measurable quantity just like a major-league pitcher's fastball?

Yes, it is a measurable quantity—a distance measured with a meter stick (or its equivalent) divided by a time measured on a clock. However, when you measure the distance traveled by light in a vacuum with a properly calibrated and accurate meter stick and then measure the time of travel with a properly calibrated and accurate clock, no matter where you are in this universe or any other, no matter how fast you are moving with respect to the source of the light you are measuring, you will get exactly 299, 792, 458 meters per second. If you don't and your results are independently verified, then you will become famous as the person who finally showed Einstein was wrong. So far, hundreds have made such a claim only to be proven wrong themselves by the data.

The reason this may seem intuitively strange to you is that you probably have the notion that distance and time are inherent, independent properties of the universe. Einstein showed they are not. Furthermore, as I keep repeating, they are human inventions—quantities that we use in describing events that we observe in terms of models in which we objectively quantify our intuitive notions of space and time.

So, when some author, be he a theologian, Christian apologist, or physicist, tells you that the speed of light is fine-tuned for life, he hasn't glanced at an introductory physics textbook published since early in the Reagan administration.

Now, it is important to realize that this definition of the meter is not merely a definition of a unit of length. It is much more fundamental than that. It is the definition of what we mean, in our models, by spatial interval—defined by the distance between two points in space. By defining space as well as time in terms of a clock measurement, we are adopting a model in which space and time are not independent properties. That's why we can talk about four-dimensional space-time. Time is just another dimension we add to the three dimensions of space.

In some of the calculations I will be making in this book, I will set $c = 1$ so that the units of distance will be inherently the same as the units of time. In one common usage in astronomy, we measure time and distance in years but call the distance unit "light-years" to avoid confusion. The speed of light is then 1 light-year per year. Setting $c = 1$ gets rid of a lot of c's in equations, not only making them simpler but helping to emphasize that the value of c has no fundamental role in physics.

Now, it is true that light does not always move at c. An electromagnetic wave has a *phase velocity* c in a vacuum. If it travels in a medium it will have a phase velocity c/n, where n is the *index of refraction* of the medium. Normally, light slows down in a medium. To see how that happens, look at figure 3.1. Although the photon, the particle of light, continues to move at the speed c in the vacuum between particles in the medium, the scattering of the photon off those particles results in a zigzag path and a lower effective speed.

If a wave is not a pure sine wave but has some other shape, such as a pulse, then it is a "wavepacket" composed of many frequencies. The pulse moves as a whole at a speed called the *group velocity*. (This switching between the particle and wave picture is characteristic of quantum mechanics and will be justified later in the book.) Different frequencies will propagate with different phase velocities in dispersive media. It is possible for $v_g > c$. But c is still c. This occurs in *anomalous dispersion* when the wave packet center in the medium races ahead faster than the wave packet center in a vacuum. However, the leading edge of the wave packet in the anomalous dispersion material never

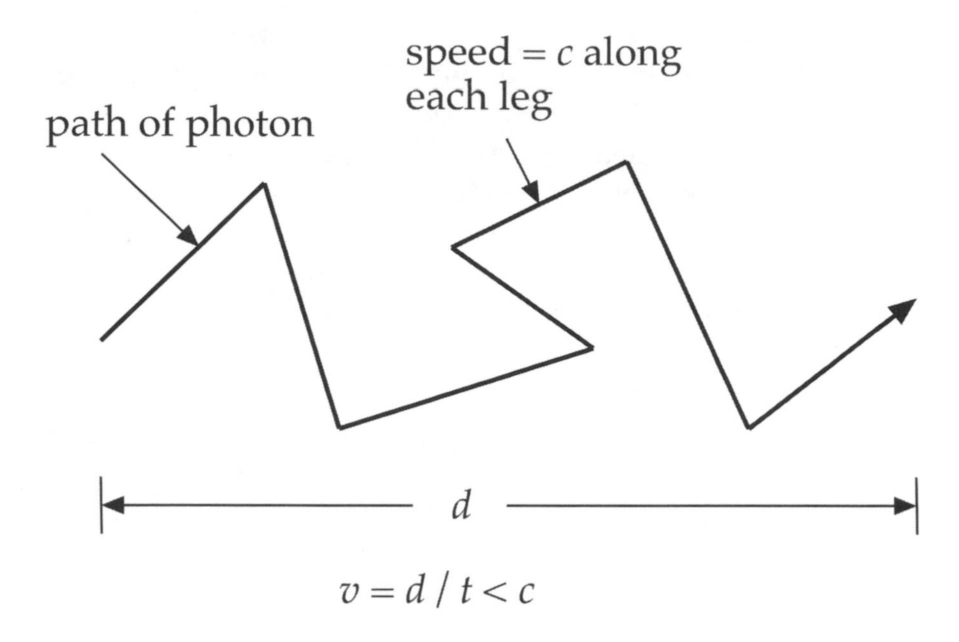

Fig. 3.1. The path of a photon in a medium of discrete particles is a zigzag as the photon scatters off the particles. While it travels as a speed c in the vacuum between particles, its net speed through the medium is less than c. If the particles are sufficiently massive and the scattering is elastic, the photon will not lose energy.

exceeds the speed of light in a vacuum. Hence the knowledge of the arrival of the wave packet, the information transfer, never violates the theory of relativity, which insists that no body or informational signal can be accelerated past the speed of light.[4]

The only particles that travel exactly at the speed c are particles with exactly zero mass, such as the photon. The photon's mass is known by the results of many observations to be less than 10^{-49} grams, which is close enough to zero for any practical application. Furthermore, the mass of the photon is exactly zero according to a basic property known as *gauge invariance* that, as we will see later in this book, is built into modern particle physics. In short, we can for all practical purposes assume that light in a vacuum travels at c.

Since the measuring unit against which everything is calibrated is the second, every physics quantity is ultimately defined by a clock measurement. If a meter stick is used to measure distance, or a thermometer is used to measure temperature, they are in principle calibrated by clock measurements ultimately, tracing back to those atomic clocks at the National Institute of Standards and Technology in Boulder, Colorado. Note that only the second is independently defined in terms of a specific measurement.

3.6. TRIVIAL AND ARBITRARY PARAMETERS

Many of the quantities that are claimed to be fine-tuned just appear so because of the units used. For example, the mass of the electron is 0.911×10^{-30} kilogram. The difference between the neutron and proton masses is 2.30×10^{-30} kilogram. If the electron were heavier by 1.39×10^{-30} kilogram, neutrons would not decay and there would be no hydrogen in the universe. Expressed this way, it naively looks as if the mass of the electron is fine-tuned to one part in 10^{30}. But actually, as we will see in chapter 10, the electron's mass could be as low as we want or 150 percent higher and we would still have hydrogen.

Three of the parameters that appear in most lists of fine-tuned quantities are arbitrary parameters:

- The "speed of light," c
- Planck's constant, h
- Newton's gravitational constant, G

The Speed of Light, c

As we just saw above, c is *by definition* 299,792,458 meters per second or one light-year per year. While light moves at this speed in a vacuum, c is fundamentally the speed beyond which a physical body cannot be accelerated according to Einstein's theory of special relativity. But its value is arbitrary. As we have seen, according to the current operational definitions of time and distance, the meter is defined as the distance light goes between two points in space in 1/299,792,458 second.

Planck's Constant, h

Most of Rich Deem's parameters are the same as Hugh Ross's, but he adds as number 31 the "uncertainty magnitude in the Heisenberg uncertainty principle."[5] I assume by this he means Planck's constant, h, since the uncertainty principle says that the uncertainty Δp_x in measuring a particle's x-component of momentum multiplied by the uncertainty in measuring a particle's position Δx must be greater than or equal to $h / 4\pi$.[6]

$$\Delta p \Delta x \geq \frac{h}{4\pi} \qquad (3.1)$$

Where did h come from? In 1900 Max Planck described the spectrum of light from a radiating body. His formula contained a parameter h now known as Planck's constant ($h = 6.626 \times 10^{-34}$ Joule-second). This was the first step in the development of quantum theory. The quantity h appears everywhere in quantum equations, more frequently in the reduced form $\hbar = h/2\pi$.

The physical meaning of Planck's constant can be obtained from the quantum theory of the atom devised by Niels Bohr in 1913. Bohr hypothesized that the only allowed electron orbits in an atom are those in which the angular momentum of the atom is an integral multiple of $\hbar$, which he called the "quantum of action."

If m_e is the mass of the electron, v is its speed, and r is its orbital radius (assumed to be a circle for simplicity in the earliest models),

$$m_e v r = n\hbar \text{ where } n = 1, 2, 3, \ldots \qquad (3.2)$$

From this Bohr was able to derive the observed spectrum of light emitted by hydrogen atoms excited by electrical sparks.

The units of $\hbar$ are those of angular momentum. If we work in units where $\hbar = 1$, then the basic quantum of action is dimensionless and $\hbar$ disappears from all our equations in the same way that c does when we set $c = 1$. In that case momentum is measured in units reciprocal to the units of distance. So, again, the value of h is arbitrary and depends on our choice of units.

Newton's Constant, G

Newton's law of gravity says that the gravitational force between two bodies is proportional to the product of their masses and inversely proportional to the distance between their centers. The constant of proportionality, G, is known as *Newton's constant*. The value of G was unknown to Newton and was first measured by Henry Cavendish in 1798. Its current best value is 6.674×10^{-11} m^3 kg^{-1} s^{-2}.

Suppose we have particles of mass m_1 and m_2 separated by a distance r. Then the gravitational force between them is

$$F = G\frac{m_1 m_2}{r^2} \tag{3.3}$$

By using the calculus he invented—simultaneously with but independently of Gottfried Leibniz (d. 1716)—Newton was able to show that this relation held for two separated bodies of arbitrary size and shape, where r is the distance between the bodies' centers of gravity.

The value of G, like c and h, depends on the system of units being used and likewise is not a universal constant. This is because the unit of mass is arbitrary and G will depend on the choice of units.

If we set $G = \hbar = c = 1$, we have what are called *Planck units*, which are commonly used in advanced cosmological research. We will find that the only parameters that may legitimately be considered in the fine-tuning question are those that are "dimensionless," that is, those that have the same value regardless of the system of units being used. However, to avoid being too confusing, I will still refer to the masses of particles in familiar units such as kilograms, with the understanding that the calculations could have been done with Planck units in which mass is dimensionless, with the same conclusions.

3.7. SPACE-TIME AND FOUR-MOMENTUM

In this and the following sections I will present a modern version of basic physical principles, including Newtonian physics, stated in the elegant formulation of four-dimensional space-time and its conjugate four-dimensional energy-momentum or *four-momentum*. And I will show that these principles are not fine-tuned since they can hardly be anything else. Some of my conclusions will be unfamiliar even to physicists, but I ask them to take a look. All the physics I use is completely well established, with only few small, conservative extrapolations beyond what is already known with great confidence. I make no unfounded speculations whatsoever. I am just proposing a simpler, less metaphysical interpretation of the principles and laws of physics than is conventional. I am not changing those laws.

A physical event is defined by its position in three-dimensional space and its time of occurrence. We can think of time as a fourth (or, more conventionally today, zeroth) coordinate in a four-dimensional space-time. The position of a particle in space-time will be designated as a four-vector $x = (x_0, x_1, x_2, x_3)$. The last three components form the familiar set of three-dimensional Cartesian coordinates measured with a meter stick that form the three-vector position, a vector in three-dimensional space: $\mathbf{r} = (x, y, z)$. I follow the convention of representing three-vectors in boldface. It will suffice for our purposes to limit ourselves to Cartesian coordinates.

The "zeroth" component, x_0, of the space-time four-vector is the time dimension, whose exact definition depends on convention. The simplest convention is to define $x_0 = ict$, where t is the time, c is the speed of light in a vacuum, and $i = \sqrt{-1}$, the unit *imaginary number*. The position of a point s in space-time with this convention is illustrated in figure 3.2.

The geometry of a space of any dimension is specified by a mathematical object called the *metric*. It tells us how to calculate the distance between two points in that space or space-time. With the choice of $x_0 = ict$, space-time is Euclidean and we can calculate space-time distances in the familiar way we do in Euclidean three-dimensional space without worrying about the metric. For example, the space-time length s in figure. 3.2 is given by the Pythagorean theorem.

In the case of three-dimensional space, an infinitesimal displacement ds is given by

$$ds^2 = g_{ij} dx_i dx_j \tag{3.4}$$

where we use the convention that repeated Latin indices are summed from 1 to 3. If the space is Euclidean, the metric is

$$g_{ij} = \begin{pmatrix} 1 & 0 & 0 \\ 0 & 1 & 0 \\ 0 & 0 & 1 \end{pmatrix} \tag{3.5}$$

and

$$ds^2 = (dx_1)^2 + (dx_2)^2 + (dx_3)^2 \tag{3.6}$$

In the *Minkowski* four-dimensional space-time used in special relativity various metrics can be used. The most common convention used by the professionals today is

$$g_{\mu\nu} = \begin{pmatrix} 1 & 0 & 0 & 0 \\ 0 & -1 & 0 & 0 \\ 0 & 0 & -1 & 0 \\ 0 & 0 & 0 & -1 \end{pmatrix} \tag{3.7}$$

So $x_0 = ct$ and the *proper time* τ is given by

$$(c\tau)^2 = g_{\mu\nu} x_\mu x_\nu = (ct)^2 - x^2 - y^2 - z^2 \tag{3.8}$$

where repeated Greek indices are understood to be summed from 0 to 3. The proper time is the time measured on a clock at rest in the reference frame where $x = y = z = 0$.

Note that, with this convention, the metric is non-Euclidean, even in special relativity. As we will see, in general relativity the metric $g_{\mu\nu}$ will depend on position in space-time.

For pedagogical purposes in the beginning I am going to use a Euclidean metric for space-time and define $x_0 = ict$ where $i = \sqrt{-1}$, the unit *imaginary number*. The metric will be

$$g_{\mu\nu} = \begin{pmatrix} -1 & 0 & 0 & 0 \\ 0 & -1 & 0 & 0 \\ 0 & 0 & -1 & 0 \\ 0 & 0 & 0 & -1 \end{pmatrix} \tag{3.9}$$

The proper time will still be given by (3.8) .

The position of the event in space-time is specified by a 4-dimensional vector ("four-vector"), as illustrated in figure 3.2

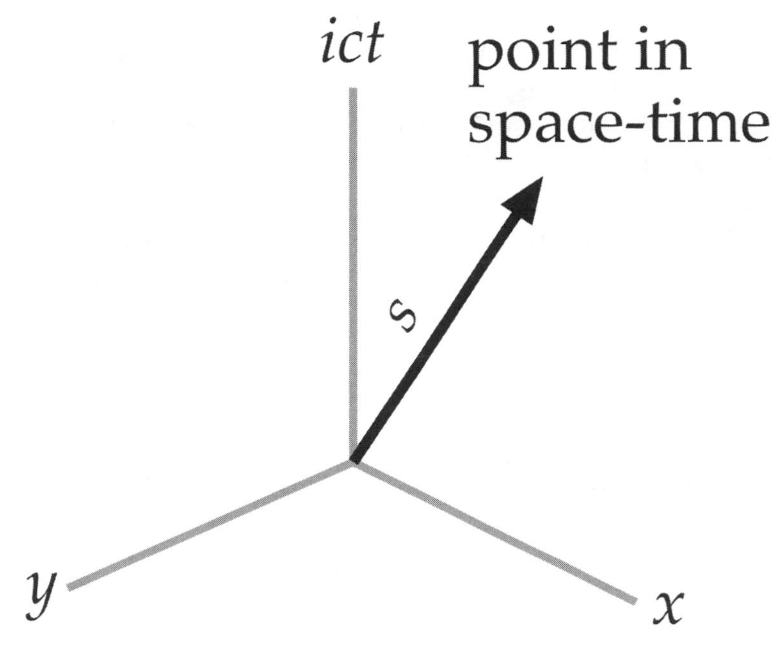

Fig. 3.2. The position of an event is a point in space-time. Shown are three of the four coordinate axes. The magnitude of the position vector is the proper distance from the origin, or c times the proper time.

with the z-axis suppressed. The magnitude of the position vector, the *proper distance*, is given by the Pythagorean theorem:

$$\begin{aligned} s^2 &= (ict)^2 + x^2 + y^2 + z^2 \\ &= -(ct)^2 + x^2 + y^2 + z^2 \\ &= -(c\tau)^2 \end{aligned} \qquad (3.10)$$

Next we need to define the *four-momentum*. In classical mechanics it is shown that for every spatial coordinate x there exists a *canonical* momentum p_x. In relativity we define a four-dimensional momentum with a zeroth component that is canonical to the zeroth component of space-time:

$p = (p_0, p_1, p_2, p_3)$. The last three components form the conventional three-momentum, a vector in three-dimensional space: $\mathbf{p} = (p_x, p_y, p_z)$, where again the boldface notation for three-vectors is used.

I will occasionally refer to **p** as "linear momentum" to distinguish it from angular momentum (see next chapter).

For now I will use the convention that the zeroth component of the four-momentum $p_0 = iE / c$ where E is the energy.

Let us define a quantity m by

$$m^2 = g_{\mu\nu} p_\mu p_\nu = \left(\frac{E}{c}\right)^2 - p_x^2 - p_y^2 - p_z^2 = \left(\frac{E}{c}\right)^2 - p^2 \qquad (3.11)$$

If you think of the four-momentum as a vector in four-dimensional momentum space, as shown in figure 3.3, the length of that vector is the *mass* of the particle m. Since the four-space is Euclidean, it can be calculated using the Pythagorean theorem. When the particle is at rest, the momentum is zero and the vector points along the iE/c axis and $E = mc^2$. In textbooks, m is usually referred to as the *rest mass*, and mc^2 the *rest energy*. In practice, physicists just call m the mass of the particle and have discarded anachronisms such as "rest mass" and "relativistic mass," except for pedagogical purposes (which probably confuse more than they enlighten).

So far everything has been completely relativistic, that is, it applies at all speeds up to the speed of light, but will still apply to Newtonian physics by just taking any speed $v \ll c$. The physics that will be used in this book is mostly Newtonian, although we will have occasion to get into special relativity, general relativity, and quantum mechanics. Even in that case I will often use what is called the *semi-Newtonian* approximation where relativistic and quantum effects are used to modify Newtonian equations. While purists will argue that this approximation is not the whole story and things are more complicated

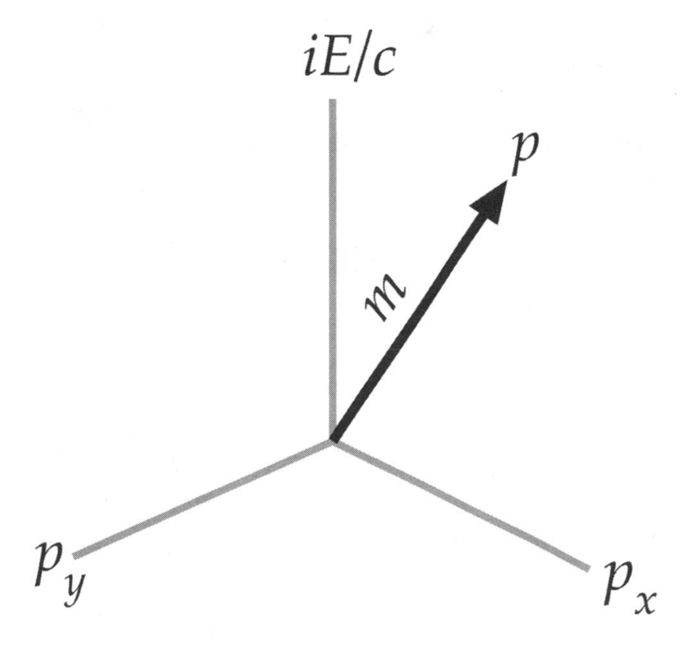

Fig. 3.3. The four-vector momentum p in momentum space. Three of four dimensions are shown. The magnitude of the four-momentum is the mass m.

than I make them out to be, I am not in the process of proposing anything new. In particular, I will repeat time and again that I do not have the burden of proving with absolute certainty every explanation I provide for some fine-tuning claim. If I provide a plausible explanation consistent with our best knowledge, then the proponent of fine-tuning, who is, after all, claiming a miracle, has the burden of proving me wrong.

3.8. CRITERIA FOR FINE-TUNING

Before analyzing specific examples of claimed fine-tuning, I need to say something about how we can decide whether or not a parameter is fine-tuned. As we have already seen, the fine-tuning that is claimed by theists is one of such incredible precision that only a supernatural being with vast powers could have brought it about. Let us look at specific examples. Without reference to any published calculation, Rich Deem claims that the *maximum* deviation from the accepted values of five parameters listed in table 3.1 would "either prevent the universe from existing now, not having matter, or be unsuitable for any form of life."[7] I have made one change to his list. I am sure he meant mass density rather than mass for the fourth parameter, which corresponds to Ross's selections.

So what do I have to do to convince a disinterested observer that a parameter is not fine-tuned? It will depend on the nature of the parameter. In the case of parameters such as those in table 3.1, if I could show that the maximum deviation is something like 10/1, or even 1/10 rather than $1/10^{37}$ or less, I expect I should be casting some doubt on the fine-tuning claim.

Table 3.1. Fine-Tuning of Five Physical Parameters

Parameter	**Max. Deviation**
Ratio of Electrons to Protons	$1/10^{37}$
Ratio of Electromagnetic Force to Gravity	$1/10^{40}$
Expansion Rate of Universe	$1/10^{55}$
Mass Density of Universe	$1/10^{59}$
Cosmological Constant	$1/10^{120}$

I need not keep reminding you that these parameters are ingredients in human-invented models and while they have something to do with reality, we know not what. Besides those parameters I have already discussed, which are trivial or arbitrary, the parameters that authors have promoted as fine-tuned fall into two further categories:

(1) **Fundamental physics parameters.** These are quantities in our physical models whose values are not provided in existing models but must be measured either directly or

indirectly. These must be regarded as variable unless and until a new model comes along that is able to calculate one or more from some other set of basic parameters. For my purposes, the main examples are the masses of the elementary particles and the strengths of the forces by which they interact. As we will see, the standard model of elementary particles and forces contains twenty-six such parameters (they are listed in chapter 4), but only a few determine the gross properties of matter or the structure of the universe. We will also see that not all these parameters are independent, as is usually assumed by fine-tuners.

(2) Fundamental cosmological parameters. These are quantities in our cosmological models that are not provided as part of existing models but also must be measured either directly or indirectly. They, too, must be regarded as variable unless and until a new model comes along that is able to calculate one or more from some other, more fundamental set of parameters. Some examples are the critical density of the universe and the cosmological constant. The concordance model of cosmology contains eleven adjustable parameters (they are listed in chapter 11). Also, here we will find that the parameters are not all independent, as is usually assumed in fine-tuning arguments.

Now, you might ask, what about a universe with a different set of "laws"? There is not much we can say about such a universe, nor do we need to. Not knowing what any of the parameters are, no one can claim that any are fine-tuned.

We can think of the parameters as coordinate axes in an abstract *parameter space* (or phase space) of *n* dimensions. The set of thirty-seven values that fits the data for our universe is then a point in parameter space.[8] The claim of fine-tuning is that you can consider a subspace of as many as thirty or more dimensions in which you cannot move the point that represents those thirty values. That's the maximum claimed. (Craig says fifty but provided no list.) From the serious literature on the subject I have determined that there are actually twenty-two parameters that need to be considered, although most have little bearing on the gross nature of the universe. Not all of them are independent, as I need to keep emphasizing, but it is convenient to treat these separately as long as their correlation is taken into account.

Note that in all the examples of fine-tuning given in the theist literature, such as the lists of Ross and Deem, the authors only vary one parameter while holding all the rest constant. This is both dubious and scientifically shoddy. As we will see in several specific cases, changing one or more other parameters can often compensate for the one that is changed. There usually is a significant region of parameter space around which the point representing a given universe can be moved and still have *some* form of life possible.

Before discussing specific examples, let me illustrate the general idea. Let us look, for simplicity, at the toy two-parameter space illustrated in figure 3.4. There we have the two axes *Par1* and *Par2*. The dashed box shows the possible parameter space allowed by the model for any universe. While that space will usually be limited, let us allow for the possibility that it is infinite. That is, the parameters can take on any value from $-\infty$ to $+\infty$, where by infinity here we do not mean a specific number but rather that the range has no limit. The dot gives the value for our universe. The wedge-shaped area, bounded by solid lines, gives the range of parameters that allow for some form of life. The area outside the wedge, A_1, contains nonviable universes, while the area of the wedge, A_2, contains the viable points. Assuming a uniform distribution of points, the probability for finding a universe with life is the ratio of the areas, $A_2/(A_1+A_2)$.

In the fine-tuning view, there is no wedge and the point has

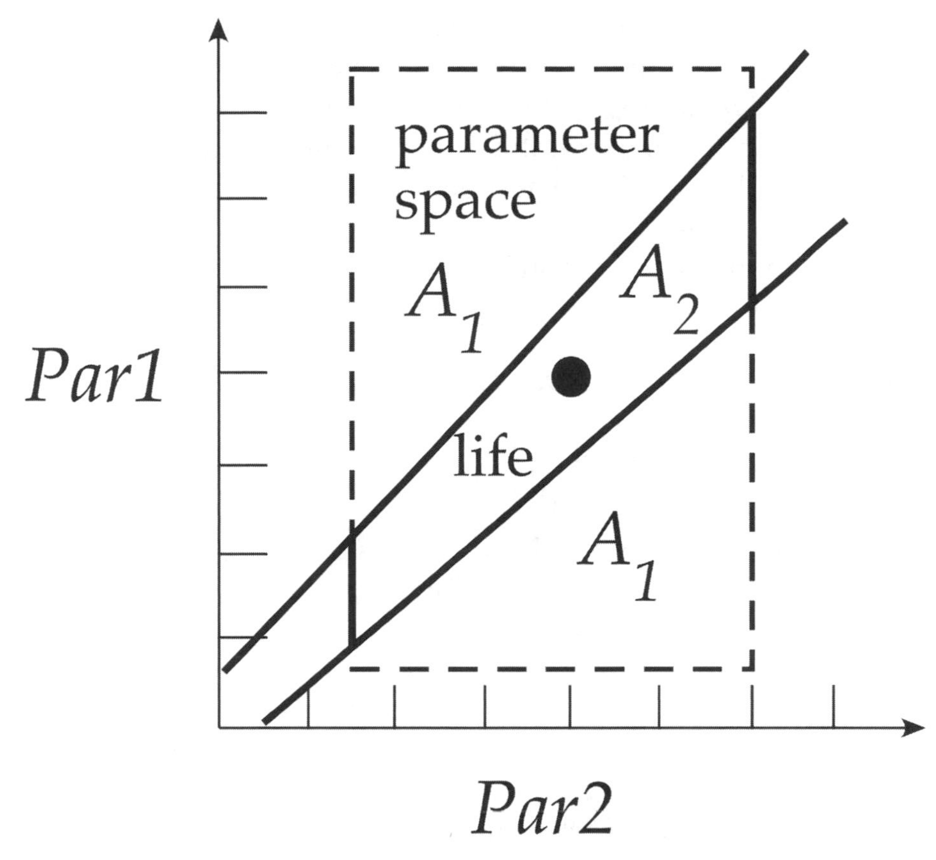

Fig. 3.4. An illustration of a two-parameter space in which the wedge-shaped area is the region where life is possible. The dashed area is the range allowed by the model, and can be infinite. The ratio of areas $A_2/(A_1+A_2)$ gives the probability for finding life. The dot represents our universe.

infinitesimal area, so the probability of finding life is zero—without God poking his finger at that point.

Now, it would seem that the probability is also zero when the parameter space is unlimited and A_1 is infinite. However, in that case, the wedge area A_2 is also infinite. The probability can then be estimated by just taking the ratio of areas in the quadrant. Even a probability of a few percent undermines the fine-tuning hypothesis, which rests on probability numbers more like 1 in 10^{120}.

Unfortunately, we have no way of knowing that the distribution is uniform, but it's the best we can do. Thus the probabilities we arrive at are not to be taken literally but rather to be treated as a "figure of merit." But since the fine-tuning claims usually involve infinitesimal probabilities like one part in 10^{123}, all I have to do to undermine each claim is to show that a natural explanation can be found with some reasonable likelihood.

3.9. GEOLOGICAL AND BIOLOGICAL PARAMETERS

Ross, Deem, and other proponents of fine-tuning also list a large number of properties that a planet must have to support biological life. For example, Ross lists the levels of oxygen, nitrogen, carbon dioxide, water vapor, and ozone in the atmosphere. No doubt earthly biological life depends on these. He also lists tectonic plate activity, oceans-to-continent ratio, soil mineralization, thickness of crust, axial tilt, gravity of the moon, and many others.[9] In other words, if a planet is not exactly like Earth in every respect, then it will not have life.

Now, I agree that earthly life is only possible on Earth and humanity will never find another planet in the universe on which it can live without massive life support. But I have a different view of life than Ross and his fellow Christians who seem to believe that we are made in God's image, presumably with God's DNA, and so no other form of life is possible.

In my view, life is a property that any sufficiently complex, nonlinear, interacting, dissipative system will develop in a sufficiently long time. So I will ignore those parameters that constrain life to our biology and our biology alone. At the same time, I realize that I cannot open the door so wide to let every wild speculation through, so my position will be middle ground. I will use what we know about life on this lonely planet and see how far the parameters of nature can be stretched to allow life of some form that is not too distant from our own.

4.

Point-of-View Invariance

4.1. THE CONSERVATION PRINCIPLES

In the previous chapter we talked about various definitions used in physics. Now let us get to the principles. In 1916, an obscure German mathematician named Emmy Noether proved one of the most important theorems in the history of physics. She showed that *for every differentiable symmetry there exists a conservation law*. This is known as *Noether's theorem*.[1]

When a physical system is unchanged under some operation, it is said to be *symmetrical* to that operation. For example, a sphere can be rotated about any axis by any amount and it looks the same, so it has *spherical symmetry*. A long line of telephone poles alongside a highway has *space-translation symmetry*, since the line looks unchanged as you drive along. Another term we will use as synonymous with symmetry is *invariance*. The sphere possesses rotational invariance. The highway possesses space-translation invariance.

Suppose that the description of the motion of a particle in a certain model is unchanged when a coordinate of the particle is translated along that coordinate axis by some amount. For example, let the car described above be driving along the highway at a constant speed of 100 kilometers per hour. If you filmed it against the backdrop of telephone poles, not showing any slowing down or speeding up, you would not be able to tell from viewing the film where the car was along the highway—how far from the next town—at any given time. We say that the car has space-translation symmetry.

Noether's theorem says that the momentum *canonical* to the coordinate with space-translation symmetry will be *conserved*, that is, will not change with time. When the translation is along the x-axis, then the momentum component p_x is conserved. More generally, when the motion of the particle does not depend on any special position in three-space, the three-vector momentum **p** and its individual components is conserved.

The translation of a coordinate is equivalent to changing the location along the axis of the origin of the coordinate system. Consider when the translation is along the time axis in four-space. Then Noether's theorem says that whenever the motion of a particle does not depend on a special moment in time, the energy of the particle is conserved.

A third principle, conservation of angular momentum, follows when no special *direction* in space is singled out. When a system does not depend on any particular direction in the xy-plane, that is, rotation about the z-axis by any angle, then the angular momentum component along the z-axis L_z is conserved. The same goes for the x- and y-components L_x and L_y along their respective axes.

The conservation principles apply not just to a single particle but also to any system of particles, where the coordinates refer to the location of the center-of-mass of the system and where the linear momentum, angular momentum, and energy refer to the sum over all the particles in the system.

In short, the great conservation principles of classical physics—conservation of energy, conservation of linear momentum, and conservation of angular momentum—follow simply from the symmetries of space-time. Energy and the two types of momentum conservation are the most important principles of classical physics and continue to apply with equal importance in modern physics.

4.2. LORENTZ INVARIANCE

Like the great conservation laws, special relativity is also based on a symmetry or invariance principle. That principle is called *Lorentz invariance* and was proposed by Hendrik Lorentz in 1899. When, in 1905, Einstein applied it to special relativity, it produced a dramatic change in our understanding of space, time, mass, and energy. It demonstrated that measured time and space intervals are relative, that rest mass and rest energy are equivalent, that the speed of light in a vacuum c is invariant, and that no body can be accelerated past the speed of light.

Let (ict, x, y, z) be a set of measurements made for the space-time position of an event by observer A, with respect to some arbitrary origin (0,0,0,0). The Lorentz transformation tells you how to calculate the set of measurements (ict', x', y', z') that would be made by another observer B moving at some speed v with respect to A along the x-axis (any axis can be chosen).

$$\begin{aligned} x' &= \gamma(x - vt) \\ y' &= y \\ z' &= z \\ t' &= \gamma\left(t - \frac{v}{c^2}x\right) \end{aligned} \tag{4.1}$$

where

$$\gamma = \frac{1}{\sqrt{1 - \frac{v^2}{c^2}}} \tag{4.2}$$

is called the *Lorentz factor*. Note the observers will measure different times. A little algebra will easily show that the proper time τ is *Lorentz invariant*.

$$(c\tau)^2 = (ct')^2 - x'^2 - y'^2 - z'^2 = (ct)^2 - x^2 - y^2 - z^2 \tag{4.3}$$

Let $(iE/c, p_x, p_y, p_z)$ be a set of energy and momentum measurements made by observer A. The Lorentz transformation also can be used to calculate the set of measurements $(iE'/c, p'_x, p'_y, p'_z)$ that would be made by observer B.

$$\begin{aligned} p'_x &= \gamma\left(p_x - \frac{v}{c^2}E\right) \\ p'_y &= p_y \\ p'_z &= p_z \\ E' &= \gamma(E - vp_x) \end{aligned} \tag{4.4}$$

The mass of the body is Lorentz invariant:

$$m^2 = \frac{E^2}{c^2} - p_x^2 - p_y^2 - p_z^2 \tag{4.5}$$

Note that when a body is at rest (momentum zero) the rest energy of the body is $E_0 = mc^2$.

4.3. CLASSICAL MECHANICS

Let us see how classical mechanics follows from these invariance principles. We begin with the laws of motion.

Newton's Laws of Motion

(1) A body at rest will stay at rest and a body in motion in a straight line at constant velocity will stay in motion unless acted on by an external force.
(2) Force is equal, by definition, to the time rate of change of linear momentum.
(3) For every reaction there is an equal and opposite reaction.

The first law follows from conservation of linear momentum. When $v << c$, the linear momentum of a particle of mass m is $\mathbf{p} = m\mathbf{v}$. If $\mathbf{p}$ is conserved, $\mathbf{v}$ will be constant. If $\mathbf{p}$ is zero, $\mathbf{v}$ will remain zero. (Note on notation: I am following the convention of using boldface for three-vectors and normal type for four-vectors.)

The second law is the definition of force. It is the agent that changes the momentum of a body. When the net force on a body is zero, its momentum is conserved. Also note that when m is a constant, the second law of motion can be written as the familiar form $\mathbf{F} = m\mathbf{a}$, where $\mathbf{a}$ is the acceleration.

The third law also follows from linear momentum conservation. Suppose particle 1 strikes particle 2, imparting a certain momentum to it in a certain time interval. Thus particle 1 applies a force to particle 2.

Now look at it from the point of view of particle 2. It strikes particle 1 imparting a momentum to it. From momentum conservation, the momentum imparted to 1 has to be balanced by the momentum imparted to the other, which is in the opposite direction. It follows that the forces are equal and opposite.

Consider the basic problem in mechanics: predicting the motion of a particle. Suppose we have a particle of fixed mass

m acted on by a force $\mathbf{F}(t)$. We know its initial velocity and position. We can then calculate the velocity and position of the particle at some later time, thus predicting the motion of the particle.

$$\mathbf{v}(t) = \mathbf{v}(0) + \int \mathbf{a}(t)dt = \mathbf{v}(0) + \frac{1}{m}\int \mathbf{F}(t)dt$$
$$\mathbf{r}(t) = \mathbf{r}(0) + \int \mathbf{v}(t)dt \tag{4.6}$$

Until quantum mechanics came along in the twentieth century with its *uncertainty principle*, physics implied that the universe was simply a giant machine, the "Newtonian World Machine," in which everything that happens is fully determined by what happened before.

Of course, we need to know the force in order to predict particle motion. For this we need, besides the conservation principles, *laws of force* such as Newton's law of gravity or the equations of electromagnetism. Rather than just write them down, as is usually done in introductory textbooks—with due attention given to the reasoning of their authors—I will show how they derive from basic principles.

An analogous set of dynamical laws holds for rotational motion, which follows from angular momentum conservation and the definition of *torque* as the time rate of change of angular momentum.

4.4. GENERAL RELATIVITY AND GRAVITY

One of the several great insights of Einstein was that gravity is indistinguishable from acceleration. Imagine a closed chamber far out in space accelerating at exactly 9.81 meters per second per second, the acceleration we experience from gravity on the surface of Earth. If you were to wake up inside that chamber, which has no windows to the outside, with no memory of how you got there, you would think you were just sitting in a room on Earth.

Now, technically, if you had some accurate measuring equipment, you would be able to detect the fact that the paths of falling objects were not converging as they would on Earth, pointing to the center of Earth. So this equivalence is "local," that is, limited to an infinitesimal volume.

Consider the law of motion $\mathbf{F} = m\mathbf{a}$ for a freely falling body in terms of a coordinate system falling along with the body. Since the body does not change position in its own reference frame, both its velocity and its acceleration in that reference frame are zero. Zero acceleration means that a freely falling body experiences no external force. An observer in that reference frame has no sense of gravity.

Next, let us consider a second coordinate system fixed to a second body such as Earth. This could be any coordinate system accelerating with respect to the first. An observer on Earth witnesses a body accelerating toward Earth and interprets it as the action of a "gravitational force."

Ask yourself this: If the gravitational force can be transformed away by going to a different reference frame, how can it be "real"?

It can't.

We see that the gravitational force is an *artifact*, a "fictitious" force just like the centrifugal and Coriolis forces. These forces are introduced in physics so that we can still write a law of motion like $\mathbf{F} = m\mathbf{a}$ in an accelerating reference frame. We sit on a spinning Earth, which puts us in an accelerated reference frame. By introducing the centrifugal and Coriolis forces we still can use all the machinery of Newtonian mechanics.

The point, though, is that if we want to make all our models point-of-view invariant, then we should not include non-invariant concepts in our basic models. We can still use the notion of a gravitational force that has proved so powerful from the time of Newton, as long as we keep in mind that it is fictitious.

Here again, Christian philosopher Robin Collins misapplies physics to claim fine-tuning. He asks us to imagine what would happen if there were no gravity. "There would be no stars," he tells us.[2] Right, and there would be no universe either. However, physicists have to put gravity into any model of a universe that contains separated masses. A universe with separated masses and no gravity would violate point-of-view invariance.

While point-of-view invariance gives us the conservation principles and Newton's laws of motion, we still need laws of force to tell us what to put in for **F** in **F** =*m***a**. Let us recall Newton's law of gravity from chapter 3. From observations Newton inferred that the force between particles of mass m_1 and m_2 separated by a distance r is proportional to the product of their masses and inversely proportional to the distance.

$$F = G\frac{m_1 m_2}{r^2} \tag{4.7}$$

Einstein was able to derive Newton's law of gravity from his general theory of relativity, which went beyond Newton's theory in predicting tiny gravitational effects. These had not been part of that theory but were eventually confirmed empirically.

General relativity also relies on invariance principles. The Newtonian gravitational force on a particle in the vicinity of a large massive body depends on the mass density of the body, that is, the mass per unit volume. Density is not Lorentz invariant. It has a different value in different reference frames because a volume element is not Lorentz invariant. Einstein sought to find equations to describe gravity that were manifestly Lorentz invariant.

Einstein deduced that the source of gravity had to be represented by a mathematical object called a *tensor*. A four-vector such as the four-momentum is a rank-1 tensor, having four components and written as p_μ where a single index runs from 0 to 3. In general, any dimension is possible, but here we are working in four-dimensional space-time. A rank-2 tensor in four-dimensional space-time is written $T_{\mu\nu}$, where both indices run from 0 to 3. These tensors are all defined to be Lorentz invariant.

The specific tensor that Einstein chose for the source of gravity was the *energy-momentum tensor,* also known as the *stress-energy tensor* $T_{\mu\nu}$. You can think of $T_{\mu\nu}$ as a four-by-four matrix or table of numbers:

$$\begin{pmatrix} T_{00} & T_{01} & T_{02} & T_{03} \\ T_{10} & T_{11} & T_{12} & T_{13} \\ T_{20} & T_{21} & T_{22} & T_{23} \\ T_{30} & T_{31} & T_{32} & T_{33} \end{pmatrix}$$

The component T_{00} of the energy-momentum tensor is just the density. The other components comprise energy and momentum flows in various directions.

So Einstein hypothesized that the gravitational field is a Lorentz-invariant tensor $G_{\mu\nu}$ that is proportional to $T_{\mu\nu}$.

$$G_{\mu\nu} = -8\pi G T_{\mu\nu} \qquad (4.8)$$

where the factor is chosen so that it gives the Newtonian result for $v << c$. The two tensors each have sixteen components, so (4.8) is a shorthand for sixteen equations.

In what has become the standard model of general relativity, Einstein related $G_{\mu\nu}$ to the curvature of space-time in a non-Euclidean geometry, as described by the metric tensor $g_{\mu\nu}$. This assured that all bodies would follow a geodesic in space-time in the absence of forces, the gravitational force thus being eliminated and replaced by a curved path.

We will not have occasion to delve into the complexities of general relativity. However, it leads to Newton's law of gravity in the limit of low speeds.

The space-time symmetries we have discussed I have termed *point-of-view invariance*. That is, they are unchanged when you change reference frames or points of view. If our models are to be objective, that is, independent of any particular point of view, then they are required to have point-of-view invariance. As we have seen, when our models do not depend on a particular point or direction in space or a particular moment in time, then those models *must necessarily* contain the quantities linear momentum, angular momentum, and energy, all of which are conserved. Physicists have no choice in the matter, or else their models will be subjective, that is, will give uselessly different results for every different point of view. And so the conservation principles are not laws built into the universe or handed down by deity to govern the behavior of matter. They are principles governing the behavior of physicists.

Although total energy cannot technically be defined in general relativity and we can only talk about energy densities, both special and general relativity are at least partly consequences of Lorentz invariance. Physicists are forced to make their models Lorentz invariant so they do not depend on the particular point of view of one reference frame moving with respect to another. This was a twentieth-century upgrading of Galileo's seventeenth-century principle of relativity (which showed there was no difference between being at rest and being in motion at constant velocity) and the Copernican principle that Earth is not at rest at the center of the universe.

4.5. QUANTUM MECHANICS

In 1930, Paul Dirac wrote the classic, definitive book on quantum mechanics that is still used to this day.[3] He proposed that quantum mechanics is primarily a theory of transformations and that the operators of quantum mechanics are the generators of those transformations. Here's how he explains it in the preface to the first edition:

> The classical tradition has been to consider the world to be an association of observable objects (particles, fluids, fields, etc.) moving about according to definite laws of force, so that one could form a mental picture in space and time of the whole scheme. . . . It has become increasingly evident in recent years, however, that nature works on a different plan. Her fundamental laws do not govern the world as it appears in our mental picture without introducing irrelevancies. The formulation of these laws requires the use of the mathematics of transformations. The important things in the world appear as the invariants . . . of these transformations.[4]

Dirac described the state of a system as a vector in an abstract space. In fact, he mentions the term *wave function* only once in the whole book, in a dismissive footnote:

> The reason for this name [wave function] is that in the early days of quantum mechanics all the examples of these functions were in the form of waves. The name is not a descriptive one from the point of view of the modern general theory.[5]

Dirac introduced the symbols $|\psi\rangle$ "bra" and $\langle\psi|$ "ket" to represent the state vector and its dual (sort of like a complex conjugate) that have become standard in physics. I will use the term *state vector* but will keep things simple by just calling it ψ.

An example of a state vector in two dimensions is shown in figure 4.1 (a). In this case, ψ is a complex function with a magnitude A and phase ϕ.

We can write

$$\psi = A\exp(\phi) \tag{4.9}$$

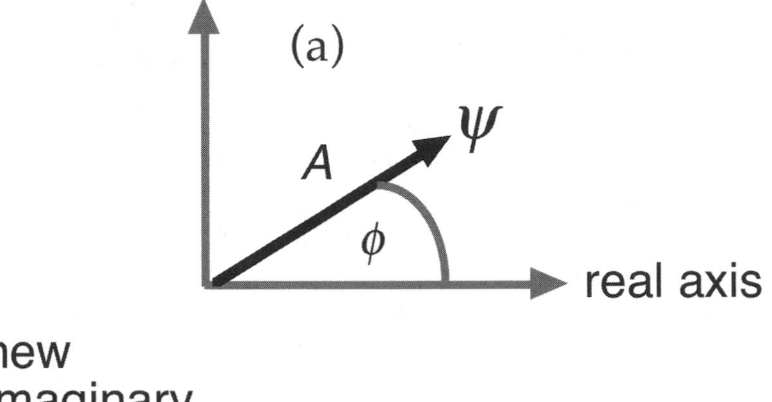

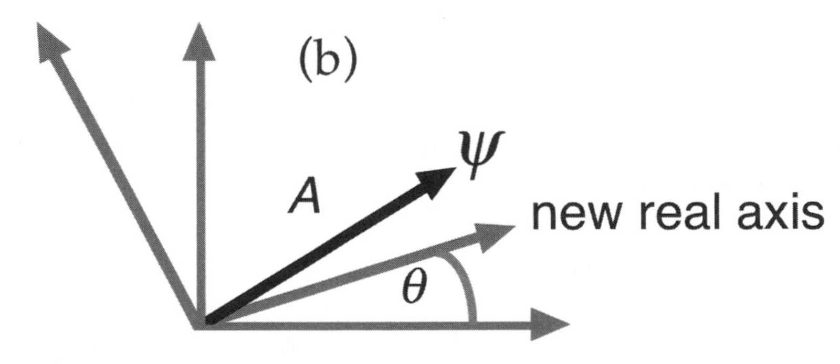

Fig. 4.1. The state vector ψ in the two-dimensional complex plane. In (a), its magnitude A and phase ϕ are shown. In (b), the coordinate system has been rotated by an angle θ. The state vector is invariant to the rotation. This is an example of a unitary transformation. Note, however, that while the vector does not change, the phase does.

In general, the state vector applies to any number of dimensions in abstract space. The *time-dependent Schrödinger equation* tells us how the state vector changes with time.

$$H\psi = i\hbar\frac{\partial\psi}{\partial t} \qquad (4.10)$$

Noether's theorem identified energy (also known as the *Hamiltonian*) as the generator of a time translation. Most physicists view this equation as the replacement of the classical **F** = *m***a**, as the equation of motion of quantum mechanics, the "law" that determines how the wave functions vary with time. But it is a trivial law that does nothing but change the origin of the time axis.

Similarly the generators of translations along the spatial axes are identified as the components of the linear momentum of the system.

$$P_x \psi = \frac{\hbar}{i} \frac{\partial \psi}{\partial x} \tag{4.11}$$

The most important transformations in quantum mechanics are called *unitary transformations*. These transformations preserve the magnitude and direction of ψ in state vector space, as seen in figure 4.1 (b). In the usual interpretation of quantum mechanics, A^2 is the probability for finding the system in the state ψ, per unit volume. "Unitarity" implies *conservation of probability*. That is, the transformation does not change the fact that the probabilities add up to one. For example, if the probability for finding a particle in a certain region of space is 25 percent, the probability for finding it outside that region is 75 percent. If we translate to another position in space where the first probability drops to 10 percent, the second will rise to 90 percent.

4.6. GAUGE THEORY

In the previous section I introduced the notion of the state vector, a vector in an abstract, multidimensional space that is used to represent the state of a system in quantum mechanics. We saw that it is invariant to transformations of the coordinate system in that space. In the simplest case, the abstract space is two-dimensional and can be represented as the complex plane where the state vector is a complex number with an amplitude A and a phase ϕ, as shown in figure 4.1(a). Rotating the coordinate system, as in figure 4.1(b), does not change A or the direction of ψ, but changes the phase ϕ. We call that a *phase transformation*. When we extend this idea to more dimensions, we have what is called more generally a *gauge transformation*. When the state vector is unchanged, the transformation is again unitary and we have *gauge invariance* or *gauge symmetry*.

As we will see, gauge invariance was the most important principle of physics discovered in the twentieth century. Note that "gauge invariance" is just a fancy technical term for point-of-view invariance. It also arises from the requirement that physicists must formulate their models in such a way that they do not depend on any subjective point of view.

Gauge invariance is a generalization of Noether's theorem from space-time to abstract state vector space. (Nothing in Noether's treatment limited it to space-time.) Consider again the two-dimensional example. In figure 4.1(b), the axes are rotated by an amount θ. The generator of the transformation is θ, which will be conserved.

Early in the twentieth century, another fact about gauge invariance was discovered. If θ is allowed to vary from point to point in space-time, Schrödinger's time-dependent equation, which we recall is the equation of motion of quantum mechanics, is not gauge invariant. However, if you insert a four-vector field into the equation and ask what that field has to be to make everything nice and gauge invariant, that field is precisely the four-vector potential that leads to Maxwell's equations of electromagnetism! That is, the electromagnetic force turns out to be a *fictitious* force, like gravity, introduced to preserve the point-of-view invariance of the system.

Note that for neutral particles, no new fields need to be introduced to preserve gauge invariance in that case.

When θ depends on the location of the particle in space-time, we have what is called a *local gauge transformation*. When it is the same at all points, we have a *global gauge transformation*. In the case of electromagnetism, θ is proportional to the electric charge of the particle and so global gauge invariance, or, as I prefer, global point-of-view invariance, leads to the principle of conservation of charge.

Also note that θ is the generator of a rotation in a two-dimensional space This generator is mathematically equivalent to angular momentum, which in quantum mechanics is quantized, that is, takes on discrete values. Thus global gauge invariance in this space will result in charge being quantized. So charge quantization is yet another consequence of point-of-view invariance.

Summarizing, we have found that the equations that describe the motion of a charged free particle are not invariant under a local gauge transformation. However, we can make them invariant by adding a term to the canonical momentum that corresponds to the four-vector potential of the electromagnetic field. Thus the electromagnetic fields are introduced to preserve local gauge symmetry. Conservation and quantization of charge follow from global gauge symmetry.

Now, we still need to introduce the *Lorentz force law*, which tells us how to calculate the force on a charged particle in the presence of electric and magnetic fields. This equation is nothing more than a definition of the fields. The electric field **E** is operationally defined as the force one measures on a charged particle at rest in the presence of other charged particles, per unit charge. More complicated, **B** is the vector field that leads to a measured force **F** on a charged particle moving with a velocity **v** that is perpendicular to the plane of **v** and **B**. How do these connect to Maxwell's equations, which are used to calculate the fields for any given charge and current distribution (a current is a moving charge)? In a 1989 paper, physicist Freeman Dyson provided a derivation of Maxwell's equations from the Lorentz force law that he says was first shown to him by Richard Feynman in 1948.[6] That is, Maxwell's equations follow from the definition of the electric and magnetic fields. If we assert that Maxwell's equations follow from point-of-view invariance, then the Lorentz law is implied.

4.7. THE STANDARD MODEL

And so we have derived all of classical physics, including classical mechanics, Newton's law of gravity, and Maxwell's equations of electromagnetism, from just one simple principle: the models of physics cannot depend on the point of view of the observer. We have also seen that special and general relativity follow from the same principle, although Einstein's specific model for general relativity depends on one or two additional assumptions. I have offered a glimpse at how quantum mechanics also arises from the same principle, although again a few other assumptions, such as the probability interpretation of the state vector, must be added.

Much of the standard model of elementary particles and forces also follows from the principle of gauge invariance. The elementary particles of the standard model are listed in table 4.1. Besides the electromagnetic force, which is described by a four-vector field A_μ, $\mu = 0,1,2,3$, there are two forces that operate only on the nuclear and subnuclear scale. The *weak nuclear force* is described by three four-vector fields $W_\mu^+, Z_\mu^0, W_\mu^-$, while the *strong nuclear force* is described by eight four-vector fields g_μ^a, $a = 1, \ldots, 8$.

Shown in table 4.1 are the three generations of spin 1/2 fermions—quarks and leptons—that constitute normal matter and the spin 1 bosons that act as force carriers. Each has an antiparticle, not shown. The top row of quarks has charge +2/3 in terms of the unit charge *e*. The second row has charge –1/3. The antiquarks have opposite charge. The neutrinos in the top line of leptons are electrically neutral. The leptons in the second line, which include the electron, have charge –1. Their antipar-

Table. 4.1. The Fundamental Particles of the Standard Model

Fermions (antiparticles not shown)				Bosons
Quarks	u	c	t	γ
	d	s	b	g
Leptons	ν_e	ν_μ	ν_τ	Z
	e	μ	τ	W

ticles have charge +1. The electroweak force is carried by four spin 1 bosons, W^+, W^-, Z, and the strong force by eight spin 1 *gluons*, g. The photon, Z, and gluons are neutral. The W charges are indicated by their superscripts. A yet unobserved neutral, spin zero *Higgs boson*, not shown, is also included in the standard model.

Each of these fields has associated with it a particle with intrinsic angular momentum (*spin*) equal to 1 in units of $\hbar$ (which is usually set equal to 1 for convenience). The particle associated with each field is termed a *gauge boson*. The gauge boson associated with the electromagnetic field A_μ is the photon, symbol γ. The weak fields $W_\mu^+, Z_\mu^0, W_\mu^-$ are associated with the weak bosons W^+, Z^0, W^-, where the superscript gives the charge. The strong fields g_μ^a are associated with eight *gluons* g^a. (Don't confuse these g's with the metric tensor.)

There are twenty-six parameters that describe the standard model:

- the electromagnetic strength α (fine-structure constant)
- the strong interaction strength α_S
- the masses of six quarks, six leptons, Higgs boson, W bosons, and Z boson
- four parameters that describe how quarks mix with one another
- four parameters that describe how neutrinos mix with one another

As we have seen, the gravitational strength is not a separate parameter but is set by the proton mass. This is a bit inconsistent, since the proton mass is itself not a fundamental parameter. However, any mass such as the electron mass could have been used.

The parameters of the standard model fields have been determined with great precision from fitting data from thirty years of high-energy particle accelerator experiments and, in the case of the cosmological constant, from astronomical data. As I write this, the Large Hadron Collider (LHC) in Geneva has produced the first collisions at its design energy in which proton and antiproton beams strike each other with a total of 7 trillion (7×10^{12}) electron volts. Already the data look interesting, deviating from some theoretical projections. When sufficient events are accumulated, we may catch a glimpse at the more fundamental fields that were assumed way back in the 1970s when the standard model was first introduced. These were fields that arose from the same assumption of local gauge invariance that give us the photon field. It will be interesting to see if gauge theory holds up with the new results.

In the standard model, the weak fields are not gauge invariant. Instead, the three different gauge fields are mixed with the photon field in a unified *electroweak* interaction that is fundamentally gauge invariant. But this symmetry is broken at lower energies by a process called *spontaneous symmetry breaking*. Although the underlying principles of the standard model are symmetric, this model only applies to the very hot universe when it was a trillionth of a second old—energies that are now finally being reached in the Geneva laboratory.

We can look at it this way. The universe we live in today is mostly very cold, only 3 degrees on the Kelvin scale or 270 degrees below zero on the Celsius scale. Of course, Earth is warmer, the sun warmer still, and the temperatures in high-energy particle collisions approach those in the early universe. We happen to enjoy the warming of the sun. In our currently cold universe, some of the symmetries are broken just as the symmetry of a sphere of water vapor is broken when it freezes into a snowflake. If our models are to be useful to us in terms of describing what we measure today, we have to break point-of-view invariance to calculate those observations we see from our own point of view, which is not universal. Even the 7 TeV energy of the LHC will not place us in the energy regime in the early universe when all the symmetries held. So we will still need a theory with broken symmetries to describe the actual measurements.

This is where we stand as we anxiously await the data from the LHC, where we will hopefully see the evidence for the next layer of human understanding of the nature of matter.

This chapter was probably tough going for the average reader. But I needed to show in some detail why the models of physics are not laws deeply ingrained in the structure of the universe and then fine-tuned by a creator to have them come out the way they did. First, they are human inventions introduced to describe observations. Second, they are just what they have to be in order not to depend on the subjective point of view of the observer. That is, they must be point-of-view invariant. This also means they will be the same in any universe where no special point of view is present. And it means that when I show in the following chapter that no laws of physics were broken for the universe to come into being, the next question, "Where did the laws of physics come from?" is already answered.

4.8. ATOMS AND THE VOID

In this, as in all my writings, I have emphasized that the best we can do in science is build models that describe previous observations and predict future observations. These models are composed of quantities that we measure according to well-prescribed procedures, such as position, time, and temperature. Their measurements are expressed as real numbers in the mathematics of the models. The models also contain abstract concepts, such as fields and wave functions, that are expressed as complex numbers, tensors, linear vectors, and other creations of the mathematical mind.

While our observations are imperfect shadows of whatever is the true reality, they do give us useful information with which we compose our models and describe nature.

Now, as I mentioned in the previous chapter, I am not going to go so far as to deny that the moon and other macroscopic objects—or even microscopic objects such as bacteria—are "real." Only at the submicroscopic, quantum level of phenomena is this an issue. But then, where do we draw the line?

In my 2000 book, *Timeless Reality*, I tried to make the case for a simple model of reality called "Atoms and the Void."[7] This model basically proposed that the universe is composed of irreducible, localized objects, "atoms," moving around in an empty void. I know, Democritus thought of it first. I am just suggesting that Democritus, Epicurus, Lucretius, and all the other figures throughout history who promoted the atomic model of matter were right all along.

Only now, twenty-four centuries after Democritus, can we fill in the details. We now have the standard model of elementary particles and forces, so we can identify our "atoms," used in that term's original meaning as "uncuttable" objects such as leptons, quarks, and gauge bosons. If the LHC or any future developments change the nature of the basic objects, this is just a detail. They could be strings or further constituents yet to be discovered. I will call them *particles* for convenience.

In this regard, my use of "atom" here should not be confused with its conventional modern use as the composite system of a tiny nucleus surrounded by a cloud of electrons that became identified with the chemical elements at the turn of the twentieth century. Originally these elements were uncuttable, since chemical reactions lacked the necessary energy to split them apart. That energy became available with the discovery of nuclear radioactivity and the eventual construction of particle accelerators, which were originally referred to as "atom smashers."

An important component of this model is time reversibility. That is, there is no arrow of time at the fundamental level, and all processes can proceed in either time direction. As I show, this solves a number of the apparent paradoxes of quantum mechanics so that no superluminal ("nonlocal") connections are required.

While I cannot prove that ultimate reality is just atoms and the void, at least it allows me to justify my assertion that the moon and bacteria are real. They are just made of particles.

5.
Cosmos

5.1. SOME BASIC COSMOLOGY

Let me review a bit of fundamental cosmology, which is based on Einstein's general theory of relativity. In general relativistic cosmology, the expansion of an assumed homogeneous and isotropic universe is described by a dimensionless scale factor $a(t)$ that characterizes the distances between bodies as a function of time. That is, if you have two bodies that are at rest with respect to the average motion of matter in their part of the universe (they are said to be "comoving" with respect to the universe), the distance between them will change with time because of the expansion of the universe. The behavior of $a(t)$ is determined by two equations derived from general relativity by the cosmologist and mathematician Alexander Friedmann in 1922.

$$\frac{1}{a^2}\left(\frac{da}{dt}\right)^2 = \frac{8\pi G\rho}{3} - \frac{kc^2}{a^2} + \frac{\Lambda c^2}{3} \quad (5.1)$$

and

$$\frac{1}{a}\frac{d^2a}{dt^2} = -\frac{4\pi G}{3}\left(\rho + \frac{3p}{c^2}\right) + \frac{\Lambda c^2}{3} \quad (5.2)$$

where ρ is the average mass density and p is the pressure of the matter and radiation in the universe. The quantity Λ is the *cosmological constant*.

In these equations the geometry of the universe is specified by a parameter k. The universe is spatially flat (Euclidean geometry) if $k = 0$; the curvature of space is positive and the universe is closed if $k = +1$; the curvature is negative and the universe is open if $k = -1$. An open universe expands forever, while a closed universe eventually recollapses. When $k = 0$, everything, on average, eventually comes to a halt.

The *concordance model* of cosmology, which I will refer to also as *the standard model of cosmology*, currently agrees with most observations.[1] The parameters of one of the latest versions will be given in chapter 11.

Almost all current cosmological models assume that the universe is divided into four different kinds of stuff:

(1) *Matter* that is composed of "nonrelativistic" bodies with $v << c$ and negligible pressure. This includes ordinary atomic or baryonic matter, both luminous as in stars and nonluminous as in planets and dust. It also includes still unidentified invisible *dark matter*. I am using the term "atomic" here to refer to the elements of the chemical periodic table.

(2) *Radiation*, which is the archaic name given to extreme relativistic bodies, that is, bodies moving at or near the speed of light. This includes the photons in the cosmic background radiation and possibly neutrinos if their masses are low enough, as discussed below. Note that radiation and matter are the same kind of stuff—

particles—that differ only in their speeds. Particles moving at intermediate speeds, high enough to be relativistic but not so high to be extreme relativistic are not a significant component of the universe.

(3) *Dark energy* with a negative pressure that provides for the current accelerated expansion of the universe.

(4) *Neutrinos* also left over from the early big bang that are virtually impossible to detect. The exact masses of neutrinos are unknown, but at least one of the three species of neutrino has a mass greater than its average kinetic energy of 2.5×10^{-4} eV corresponding to the temperature of 3 degrees Kelvin, which makes the members of that species nonrelativistic "matter." Neutrinos with rest energies much below 2.5×10^{-4} eV would be "radiation" ($v \approx c$). In the early universe, when the temperature was much higher, they would all have been extreme relativistic, and members of that species would be radiation.

Note, however, except possibly for the dark energy, everything still is composed of material bodies, including radiation. They have all the properties that define matter: inertia, mass (which can be zero), momentum, spin (intrinsic angular momentum, which can be zero), and energy. In general, the total energy of a body is composed of rest energy, kinetic energy, and, when interacting with other bodies, potential energy.

We can write the Friedmann equations in such a way that the various components are represented by an energy density.

The first Friedmann equation (5.1) becomes

$$\frac{1}{a^2}\left(\frac{da}{dt}\right)^2 = \frac{8\pi G}{3}(\rho_m + \rho_r + \rho_k + \rho_\Lambda) \qquad (5.3)$$

where ρ_m is the matter density, ρ_r is the radiation density,

$$\rho_k = -\frac{3kc^2}{8\pi Ga^2} \tag{5.4}$$

is the *curvature energy density*, and

$$\rho_\Lambda = \frac{\Lambda c^2}{8\pi G} \tag{5.5}$$

is the *cosmological constant energy density*. The second Friedmann equation (5.2) becomes

$$\frac{1}{a}\frac{d^2a}{dt^2} = -\frac{4\pi G}{3}(\rho_m + 2\rho_r - 2\rho_\Lambda) \tag{5.6}$$

where the pressure from nonrelativistic matter is negligible, but the pressure from radiation is $p_r = \rho_r c^2 / 3$ with a similar form for the cosmological constant term.

Each of the components of the universe has a *gravitational mass*, $M_i = E_i / c^2$, where E_i is its total energy. The gravitational mass is equal to the inertial mass by the equivalence principle. The relative contribution of each component to the total gravitational mass of the universe is given in table 5.1.

Table 5.1. The Gravitational Mass/ Energy Budget of the Universe

Component	Percentage
Radiation and Neutrinos	0.005%
Ordinary Luminous Matter	0.5%
Ordinary Nonluminous Matter	3.5%
Dark Matter	23%
Dark Energy	73%

While there are a billion times as many photons and neutrinos in the cosmic background as there are atoms in the universe, the total energy of photons and neutrons is negligible at 0.005 percent. The visible (luminous) atomic matter constitutes only 0.5 percent, and all atomic or baryonic matter only 4 percent.

The nature of dark matter has not yet been determined, although hopes are high that the Large Hadron Collider (LHC) will reveal its ingredients. They are known not to be baryonic and so must be a form of matter not yet identified. The dark matter contributes 23 percent of the mass of the universe.

The dark energy, which makes the largest contribution to the gravitational mass of the universe, is even more mysterious than the dark matter. The dark energy is probably responsible for the current acceleration of the expansion of the universe, which was first discovered in 1998. This observation, now well confirmed, first came as a great surprise. Instead of having the familiar gravitational attraction with other matter, dark energy repels it. If Earth were made of dark energy, everything would fall up!

Still, dark matter and dark energy should not be regarded as phenomena outside the realm of material physics. They, too, have all the defined properties of matter.

It turns out that when Einstein developed his general theory of relativity, he found that his theory contains an arbitrary constant that can be positive or negative. If positive, this *cosmological constant*, symbolized by Λ, provides for repulsive gravity. At the time, Einstein was unaware that the universe was expanding—a fact that was not discovered until 1929 by astronomer Edwin Hubble. So Einstein included the cosmological term to provide a stabilizing force to balance attractive gravity. After Hubble's discovery, Einstein referred to the cosmological constant as "the biggest blunder of my life," and for many years after it was assumed to be zero.

You will often hear the cosmological constant referred to as a "fudge factor" that Einstein introduced to stabilize the universe. Actually, this is misleading. Einstein's equation for general relativity necessarily contains this factor, and we need some reason, some principle, to justify setting it equal to zero. No one has yet discovered such a principle, although a highly technical new theory called *scale relativity* that unites relativity and quantum mechanics, invented by astrophysicist Laurent Nottale, predicts a nonzero value of the cosmological constant in the right ballpark.[2] Nottale's proposal, which makes a number of other remarkable predictions, is beyond the scope of this book. However, I will discuss the cosmological problem in greater detail in chapter 12.

The cosmological constant is equivalent to a constant mass density that can be added to the cosmological equations as another source of gravity, albeit repulsive in this case if the constant is positive. This can be seen from (5.2), where the second equation is just Newton's second law of motion with radiation included and the cosmological term has the opposite sign to the more familiar, attractive gravity term.

As we will see later, the cosmological constant may have been responsible for the phase of exponential expansion called *inflation* that is believed to have taken place in the early universe. Furthermore, it is currently the best explanation for the dark energy. While observations are consistent with that hypothesis, the possibility is still open for the dark energy to be composed of some form of matter with negative pressure, usually referred to as "quintessence." When the pressure of any substance is sufficiently negative, gravity will be repulsive.

5.2. A SEMI-NEWTONIAN MODEL

Let us construct a "semi-Newtonian" model in which the universe is an expanding, uniformly dense sphere, as illustrated in figure 5.1. A particle of mass m at a distance r from the center of the sphere is moving with the expansion at a speed v.

Friedmann's equation (5.6) leads to a generalization of Newton's law of gravity, which was derived in chapter 3 and was given in equation (3.3),

$$F = -\frac{Gm}{r^2}\left(M_m + 2M_r - 2M_\Lambda\right) \tag{5.7}$$

where $M_i = \dfrac{4\pi}{3}\rho_i r^3$.

The total mass of the sphere is the sum of the gravitational masses associated with each type of component of the universe: mass M_m, radiation M_r, and the cosmological constant (or, if it should turn out, quintessence) inside the sphere M_Λ. The gravitational forces from radiation and the cosmological constant are given by general relativity to be twice that of normal matter by virtue of the fact that their pressure, unlike that of nonrelativistic matter, is significant. The cosmological constant

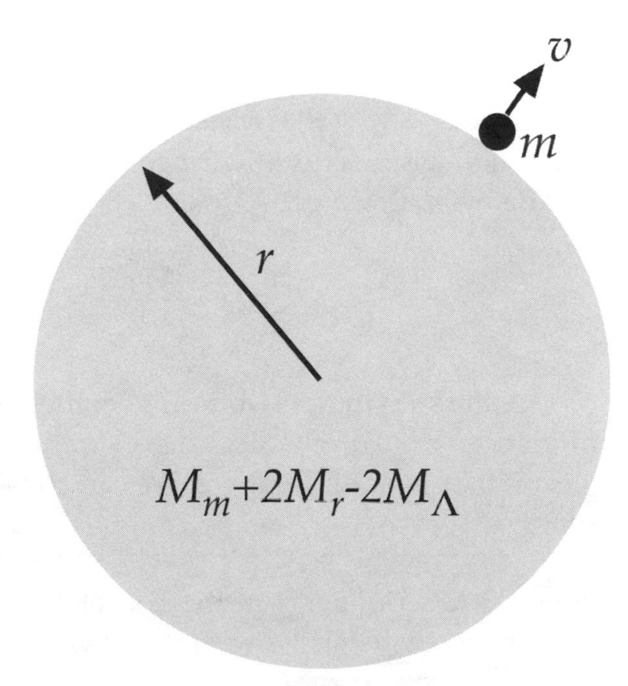

Fig. 5.1. Semi-Newtonian model of the universe. The universe is a uniformly dense, expanding sphere of matter, radiation, and dark matter. A body of mass m is at a distance r from the center and is moving with the expanding sphere at a speed v. The gravitational force on that body is given by equation (5.7).

term has the opposite sign from the others, showing that it is repulsive rather than attractive if the constant is positive.

If we neglect the small acceleration of the current expansion, we can write $v = Hr$. This is called *Hubble's law* and H is the *Hubble parameter*. When Hubble first made his discovery of the universal expansion, and for many years until the discovery of the cosmic acceleration in 1998, the data plot of v against r seemed to fit a straight line, implying a constant H. Note that this is the result you would expect from an explosion, the "big bang," if all the bodies in the explosion went off independently. The faster bodies would have gone farther in proportion to their speeds. Also note that, in that case, the current age of the universe, assuming it began with the big bang, is $T = r/v = 1/H$ evaluated now. Our current best estimate is $T = 13.7$ billion years.

Friedmann's equation (5.3) can be written, in the semi-Newtonian model,

$$v^2 = \frac{2MG}{r} \tag{5.8}$$

where $M = M_m + M_r + M_k + M_\Lambda$, just the sum of the four types of mass/energy (their pressures do not matter). Multiplying by $m/2$, we get

$$\frac{1}{2}mv^2 - \frac{GmM}{r} = 0 \tag{5.9}$$

Notice the sum of the two terms on the left is just the nonrelativistic ($v << c$) mechanical energy of the body of mass m that sits on the expanding sphere. We can consider this an average mass.

When the kinetic energy of the nonrelativistic matter of the universe is exactly balanced with its gravitational potential energy, the universe hovers between eventual collapse and eternal expansion. This occurs at the *critical density.*

$$\rho_c = \frac{3H^2}{8\pi G} \qquad (5.10)$$

Note that this does not apply just for $k = 0$, as is often thought. Curvature mass can contribute.

As we saw above, almost three-quarters of the gravitational mass of the universe is in the form of dark energy. It doesn't matter what constitutes the mass of the universe. As long as it has the critical density, galaxies and other nonrelativistic matter will have, on average, escape velocity. Observations indicate that the universe, when dark matter and dark energy are included, has an average density equal to the critical density with a small empirical error.

5.3. INFLATION, PAST AND PRESENT

In the early 1980s, three physicists, Demos Kazanas,[3] Alan Guth,[4] and Andrei Linde,[5] independently came up with an incredible suggestion about the early universe. This idea, called the *inflationary universe,* has stood the test of observations and time and is now part of standard cosmology. For a good introduction, see Guth's 1997 book *The Inflationary Universe.*[6]

Let us assume the universe has no mass or radiation, just a cosmological constant. Then, the second Friedmann equation (5.2) becomes

$$\frac{1}{a}\frac{d^2a}{dt^2} = \frac{\Lambda c^2}{3} \qquad (5.11)$$

This has the simple solution

$$a(t) = a_0 \exp\left(\frac{t}{a_0}\right) \qquad (5.12)$$

where $a_0 = a(0) = \frac{1}{c}\sqrt{\frac{3}{\Lambda}}$.

This also is a solution of the first Friedmann equation (5.1) with $k = 0$. We see that we have an exponential expansion of the universe away from $t = 0$. This is inflation. There is also a solution on the negative side of the time axis that most authors ignore. I, too, will ignore it for now but will bring it up again in chapter 6.

In the inflationary model, soon after the universe began, it underwent a short period, on the order of 10^{-32} second, of exponential expansion during which its size increased by many

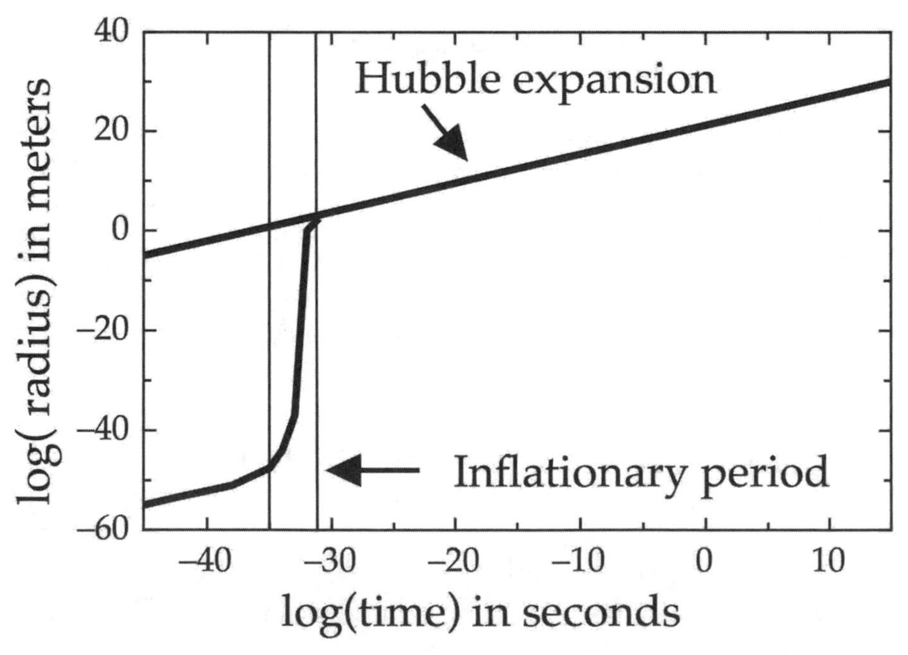

Fig. 5.2. The expansion of the early universe with the inflationary period compared with the previous big-bang model with the almost linear Hubble expansion.

orders of magnitude, at which point the more familiar big bang with its Hubble expansion took over, as illustrated in figure 5.2.[7] As we saw above, the current universe is also undergoing a much slower inflation, resulting in an acceleration of its expansion.

The easiest way to explain both epochs of cosmic acceleration is to attribute them to a cosmological constant. However, the rates of the two expansions are hugely different, so they cannot have the same constant. Furthermore, Einstein's equations would seem to imply that there is a single cosmological constant. As we will see further in chapter 12, the whole issue of the cosmological constant is seriously problematical. For now, let me just note that it is possible that Λ is zero or very small and another field or fields exist in the universe with negative pressure that account for either or both epochs of acceleration.

Now, you would not think that much can happen in 10^{-32} second, but that's the magic of exponentials. Estimates are that during inflation the size of the universe doubled every 10^{-34} second, which is one hundred doublings in 10^{-32} second. This provided simple answers to two problems with the pre-1980 big-bang cosmology: the *horizon problem* and the *flatness problem*.

The horizon problem. The observed cosmic microwave background radiation from two regions in opposite directions in the sky is characterized by the same temperature to one part in 100,000. That radiation was produced when the universe was 300,000 years old. Taking into account the expansion of the universe, those regions were 90 million light-years apart when they emitted the radiation we observed. The horizon distance at that time, according to the conventional big bang, was 900,000 light-years; the two regions were unable to interact with one another at the speed of light or less, so it would be an incredible coincidence to have the same temperature. Inflation solves the problem by increasing the size of the universe by twenty to thirty orders of magnitude so that everything in our present visible universe was causally connected.

The flatness problem. According to our best observations, the universe seems to be precisely balanced between closed and open, that is, on average, geometrically flat. The mass density of the universe is estimated to equal the critical density for perfect balance to fifty decimal places. Today this is still listed by theists as an example of fine-tuning. However, the problem was solved in 1980 by the inflationary model.

Space can be likened to the surface of a balloon that has been expanded by 20 or 30 orders of magnitude. A small patch on that surface will appear extremely flat. This also accounts for the incredible smoothness of the cosmic background radiation. Furthermore, the inflationary model predicted that the deviation from smoothness should be one part in 100,000. This prediction was spectacularly verified by the Cosmic Background Explorer (COBE) in 1992[8] and has since been verified by more advanced space-borne telescopes.[9]

While cosmologists continue to propose alternatives to the inflationary model, it is part of the current standard model of cosmology, and in this book I will not speculate too far beyond this model and the corresponding standard model of particles and forces to provide explanations for the anthropic coincidences.

Recall that the cosmological constant is not the only possible source of a gravitational repulsion that causes inflation. It can also occur with a source of matter having negative pressure we have previously dubbed *quintessence*.

5.4. PHASE SPACE

As mentioned in the preface, the state of the system is represented by a point in an abstract *phase space*. Each axis in phase space represents a degree of freedom of the system. For a single point particle in classical physics, the phase space axes are its position coordinates (x, y, z) and their canonical momentum components (p_x, p_y, p_z). For N such particles phase space has $6N$ dimensions. If instead of particles you have rigid bodies,

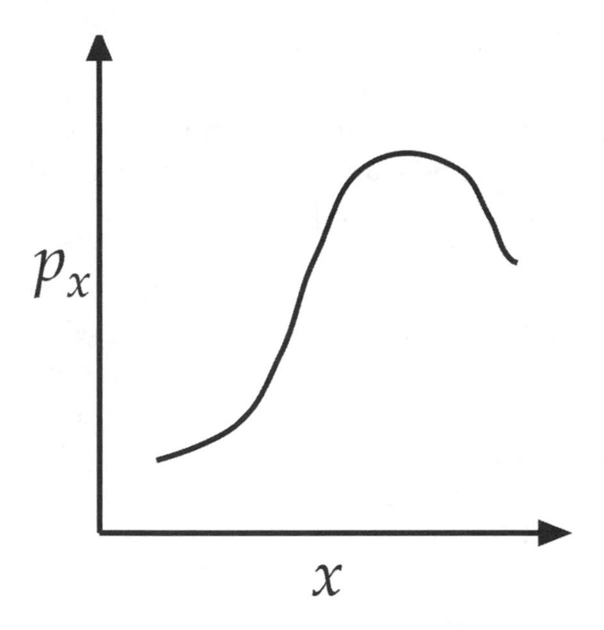

Fig. 5.3. A path in phase space for a single point particle moving in one-dimension along the x-axis.

you must add a dimension for each rotational coordinate and the corresponding components of angular momentum. If the bodies are not rigid, then you have even more dimensions to add. It can get complicated.

In figure 5.3, a two-dimensional phase space is illustrated, which corresponds to a single point particle moving along the x-axis.

5.5. ENTROPY

The notion of entropy has been playing an increasing role in cosmology. So let me introduce some of the basics. Entropy was a crucial concept in classical thermodynamics, that is, the physics of heat phenomena that was already highly developed in the nineteenth century as a practical engineering science before the introduction of the atomic model of matter. With the atomic model, thermodynamics became subsumed in statistical mechanics, in which the principles of Newtonian mechanics and, eventually, quantum mechanics were applied, with the aid of statistical methods, to many-body phenomena. All the "laws" of thermodynamics originally inferred from the observation of macroscopic solids, liquids, and gasses were derived from statistical mechanics in which each gram of matter contains 6×10^{23} elementary particles.

The most important figure in developing statistical mechanics was Ludwig Boltzmann. He made the following argument: Suppose we have a system of N particles in random motion. That system will eventually reach an equilibrium where its entropy is maximum.

Boltzmann defined a quantity H by

$$H = \sum_{i=1}^{D} P_i \ln P_i = \langle \ln P_i \rangle \quad (5.13)$$

where D is the number of degrees of freedom or dimension of phase space and P_i is the probability for finding the system of particles in the state i. He then showed that when a system is composed of randomly moving bodies, it will approach an equilibrium state where H is minimum. This is called *Boltzmann's H-theorem.*

The *entropy* of the system is defined as

$$S = -k_B H \quad (5.14)$$

where k_B is *Boltzmann's constant.* It can be shown that this is equivalent to the thermodynamic forms of entropy.

Entropy and Information

In 1948, Bell Labs mathematician and electrical engineer Claude Shannon developed the theory of information that is used today in communication engineering. He defined the information transferred to a system in a signal as the decrease in the entropy of the system.

Shannon defined a quantity we will call H_S that was essentially the same as Boltzmann's H given in equation (5.13),

$$H_S = -\sum_i P_i \log_2 P_i = -\langle \log_2 P_i \rangle \tag{5.15}$$

The only difference is he used the base-2 logarithm rather than the base-e natural or *Naperian* logarithm. This is called the *Shannon uncertainty*. Shannon then defined the information I carried by a message as the decrease in the Shannon uncertainty before and after the signal is received

$$I = H(\text{before}) - H(\text{after}) \tag{5.16}$$

Entropy and Cosmology

For the purposes in this book, the entropy of a cosmological system can be approximated by the number of particles in the system, in units where Boltzmann's constant $k_B = 1$.

Consider the case where a system has a total Ω states of equal likelihood. Then

$$S = -k_B \ln\left(\frac{1}{\Omega}\right) = k_B \ln(\Omega) \tag{5.17}$$

Further, let us assume all the particles are identical, such as all photons or all electrons. Then $\Omega = nN$ where n is the number of states accessible to a single particle and

$$S = k_B \ln\left(n^N\right) = k_B N \ln(n) \tag{5.18}$$

Now let's apply this to cosmology. First, let $k_B = 1$, so temperature is measured in energy units. Second, let us recognize that ln(n) will be a small number of order unity. So we can simply approximate $S = N$.

The maximum entropy of a sphere is equal to the entropy of a black hole of the same radius, since we have zero information about what goes on inside. Since the radius of a black hole is proportional to its mass, its entropy will increase with the square of its radius.

Consider a sphere of radius R containing particles of average energy ε. The Compton wavelength of the average energy particle will be $\lambda = hc / \varepsilon$. That particle will lose its identity when the wavelength exceeds the circumference of the sphere, so that $\lambda \leq 2\pi R$ or $\varepsilon \geq \hbar c / R$. Let the total energy in the sphere be E. Then the entropy of the sphere will be

$$S = \frac{E}{\varepsilon} \leq \frac{ER}{\hbar c} \tag{5.19}$$

Now suppose the sphere is a black hole of mass $M = E / c^2$. Its radius R will be equal or less than the Schwarzschild radius $R_S = 2GM / c^2$. So

$$S_{BH} = \frac{c^3 R^2}{2\hbar G} \tag{5.20}$$

In 1973, physicist Jacob Bekenstein published a more precise formula,[10]

$$S_{BH} = \frac{\pi c^3 R^2}{\hbar G} \tag{5.21}$$

which is close to my simple derivation. Since I ignored the ln(n) factor in (5.18), it is no surprise that my result differs slightly, but it shows the basic idea.

Let's simplify our equations by setting $\hbar = c = 1$ (natural units) and write

$$S_{BH} = \frac{\pi R^2}{G} \tag{5.22}$$

Next let us calculate the mass/energy density of a black hole

$$\rho_{BH}(R) = \frac{3M}{4\pi R^3} = \frac{3}{8\pi G R^2} \tag{5.23}$$

Suppose R equals the Hubble radius, $R_H = c/H$, that is, the radius of the visible universe. Then the average density

$$\rho_{BH}(R_H) = \frac{3H^2}{8\pi G} \tag{5.24}$$

Now, recall the first Friedmann equation, (5.3),

$$H^2 = \frac{8\pi G}{3}(\rho_m + \rho_r + \rho_k + \rho_\Lambda) = \frac{8\pi G\rho}{3} \tag{5.25}$$

where $H = \frac{1}{a}\frac{da}{dt}$

and the density ρ is the sum of all the contributions to the mass/energy of the universe: matter, radiation, curvature, and cosmological constant. From (5.10) it is also the critical density.

The average density of the visible universe is equal to that of a black hole of the same size. This does not imply, however, that the universe is a black hole, since it has no future singularity and the horizon is observer-dependent. But it does imply that the entropy of the visible universe is maximal.

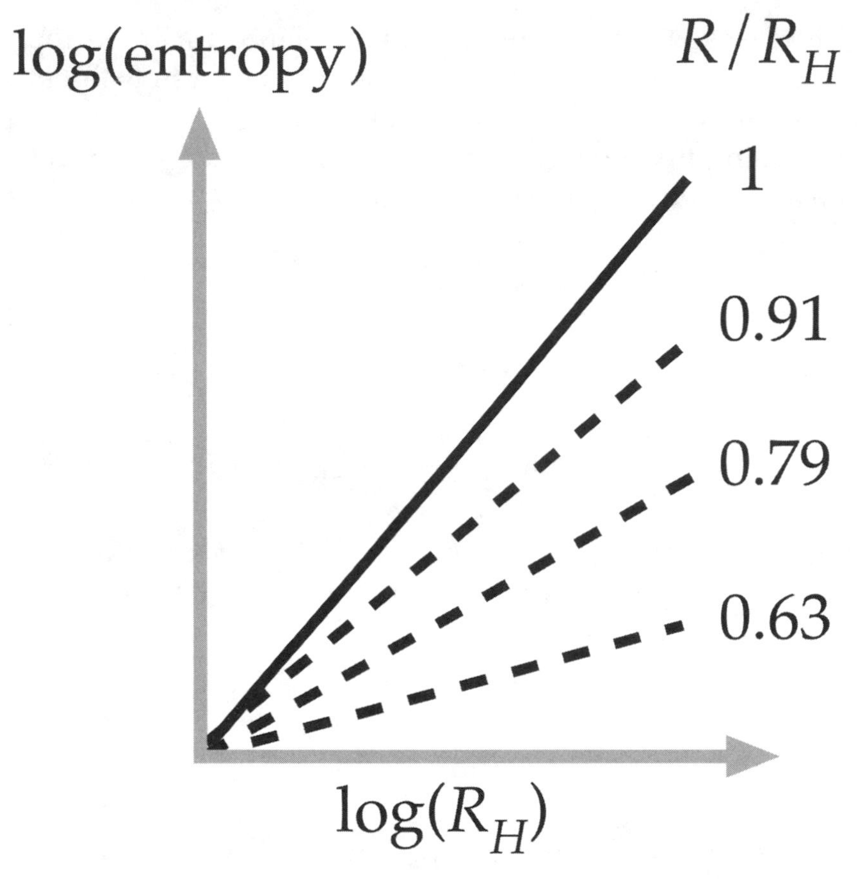

Fig. 5.4. The entropy within a sphere of radius *R* within the universe as a function of the Hubble radius, that is, the radius of the visible universe.

Now, this does not mean that the local entropy is maximal. The entropy density of the universe can be calculated. Since the universe is homogeneous, it will be the same on all scales.

$$s = \frac{S_{BH}}{V_H} = \frac{3}{4GR_H} \qquad (5.26)$$

so the entropy in a sphere of volume *V* and radius *R* is

$$S = sV = \frac{\pi R^3}{GR_H} \quad (5.27)$$

which is less than the maximum allowed for that radius, which is given by (5.22). So the difference is

$$S_{\max} - S = \frac{\pi R^2}{G}\left(1 - \frac{R}{R_H}\right) \quad (5.28)$$

As long as $R < R_H$, order can form without violating the second law of thermodynamics. In this case, the volume is less than the Hubble volume and there is room to toss out unwanted entropy. This is illustrated in figure 5.4.

Suppose we again make our semi-Newtonian approximation of the universe as a uniform, expanding sphere. Let's assume it starts out as a sphere of radius equal to the Planck length (the smallest it can be with our operational definition of space). The entropy of the universe at that time is equal to the maximum entropy of a black hole of the same radius.

From (5.27), since $R = R_H = R_{pl}$,

$$S = \frac{\pi R^3}{GR_H} = \frac{\pi R_{pl}^{\ 2}}{G} = S_{BH} \quad (5.29)$$

Thus the universe starts out with maximum entropy or *complete disorder*. It begins with zero information. It has no record of anything that may have gone on before, including the knowledge and intentions of a creator. If a creator existed, he left no record that survived that initial chaos. Once the universe exploded into the inflationary big bang, the entropy in any volume less than the Hubble volume is less than maximum, leaving room for order to form.

6.
The Eternal Universe

6.1. DID THE UNIVERSE BEGIN?

The hypothesis of a divine creation is based on a number of presumptions about cosmology. It assumes that the universe is unidirectional in time and had a beginning in time. The assumption is also made that the creator crafted certain "laws" that force matter and energy to behave in a particular way. He then meticulously set the parameters that appear in these laws when the laws are expressed mathematically so that life as we know it and, especially, humanity would eventually appear. I will challenge all these assumptions.

Certainly an all-powerful creator could have made a universe delicately balanced to produce life. But he also could have made life exist in any kind of universe whatsoever, with no delicate balancing act necessary. So if the universe is, in fact, fine-tuned to support life, it is more—not less—likely to have had a natural origin. In chapter 14 I make this argument precise using Bayesian statistics.

Certainly for a creation to have happened at all, the universe must have had a beginning. Christian philosopher and apologist William Lane Craig has led the way over the years in claiming that the universe had to have a beginning and that this necessitated a creator. He calls this the Kalâm cosmological argument,[1] which is supposedly based in Islamic sources but also has appeared in different forms in Western philosophy.[2] Recently Craig has brought his arguments up to date in a 101-page article, coauthored with US Navy warfare analyst James Sinclair.[3]

The basic argument is a syllogism. Recall my remarks about deductive logic. By itself, a deductive argument cannot tell you anything that is not already embedded in the premises. Either the premises are correct statements about nature or they are not, and the only way we have to judge them is empirically. Here is the argument:

1. Everything that begins to exist has a cause.
2. The universe began to exist.
3. Therefore the universe has a cause.[4]

We see there are two premises and a conclusion. I will grant that the logic going from the premises to the conclusion is correct, so the only issue is the truth of the two premises. Remember we cannot just look at them and say, "Yes they make sense. They agree with my own observations of the world."

Let's consider premise (1). Is it based on empirical fact? Is it a fact that everything that begins has a cause? Obviously we haven't observed the beginning of everything, so we can't say that everything that begins has a cause. As the great Scottish philosopher David Hume (d. 1776) pointed out in *An Enquiry concerning Human Understanding*, even when we observe one event following another, we cannot conclude that a causal relation between the two exist.[5] Here we have not even seen the beginning of the universe, much less a precursor event.

Furthermore, even in science, where causal relationships have been built into our models for centuries, we find that events happen without cause. When an electron in an excited energy level of an atom drops down to a lower level, it emits a photon that provides the common source of light or, more generally, electromagnetic radiation, over a spectral range from X-rays to ultraviolet light. This is a quantum phenomenon that, in most interpretations of quantum mechanics, occurs spontaneously, that is, without cause.

The radioactive decay of an atomic nucleus is another prime example. All the quantum theory of nuclei can do is predict the probability of an atomic transition or nuclear decay, and it does so quite accurately. But nothing in the theory enables us to predict when a particular atom will emit a photon or when a nucleus will decay. They are undetermined, uncaused.

Now, there is a version of quantum mechanics, proposed by physicist David Bohm in the 1950s, in which quantum events do happen deterministically. Bohm dubbed this the "ontological interpretation of quantum mechanics."[6] It is also known as the *hidden variables theory*. Although deterministic in principle, the theory itself does no better than the other versions of quantum mechanics in being able to predict only the probability of an event and not the actual outcome in every case. If one associates determinism with predictability, then Bohmian quantum mechanics is not deterministic. Furthermore, the theory requires instantaneous connections across the universe that violate Lorentz invariance, and this fact alone has made it unpopular among most physicists and philosophers.[7]

None of the interpretations of quantum mechanics provides a model that makes predictions for individual events in all cases. It then becomes a matter of how you define "cause." You can go ahead and call the quantum situation something like "indeterministic cause" if you want, but whether or not you call it "cause," the fact does not change that, to the best of our knowledge, certain events cannot be explained as the predictable result of preceding events. What is important is the probabilistic aspect. I don't think that a god who throws dice is the God that Craig and his fellow monotheistic believers worship. Craig's God created the universe with specific purposes in mind, and those purposes are hardly likely to be achieved when his actions result in random results that are not predetermined.

In any case, Craig and Sinclair cannot claim as a fact that everything that begins has a cause. Thus it follows that the Kalâm cosmological argument fails because its first premise fails.

Premise (2) also is not an established empirical fact. Craig and Sinclair base this premise on the big bang, which I agree is a well-established fact. However, nothing of what we know about the big bang requires us to assume it was the beginning of the universe, where by "universe" here I mean everything that exists in the physical realm. There are numerous theories of the origin of our visible universe in which it does not have a beginning in the sense required by the Kalâm argument. Rather, it has a beginning only in the sense that our particular observable universe arose out of a prior universe. In many of these theories, the natural process that gave rise to our universe arbitrarily gives rise to many other universes, too. I will discuss the kinds of ways in which our universe may have begun naturally in more detail later. The question here is whether it marked the beginning of time and thus the beginning of everything that is physical.

Let me quote Craig from his 1988 debate with biologist/philosopher Massimo Pigliucci, which is almost word-for-word what Craig has presented in numerous other debates:

> Have you ever asked yourself where the universe came from? Why everything exists instead of just nothing? Typically atheists have said the universe is just eternal and uncaused. But surely this is unreasonable. Just think about it a minute. If the universe is eternal and never had a beginning, that means that the number of past events in the history of the universe is infinite. But mathematicians recognize that the idea of an actually infinite number of things leads to self-contradictions. For example, what is infinity minus infinity? Well, mathematically, you get self-contradictory answers. This shows that infinity is

> just an idea in your mind, not something that exists in reality. David Hilbert, perhaps the greatest mathematician of this century, states, "The infinite is nowhere to be found in reality. It neither exists in nature nor provides a legitimate basis for rational thought. The role that remains for the infinite to play is solely that of an idea."[8] But that entails that since past events are not just ideas, but are real, the number of past events must be finite. Therefore, the series of past events can't go back forever; rather the universe must have begun to exist.[9]

Craig and Sinclair spend many pages trying to shoot down the straw man that Craig sets up in the above quotation. They prove the following syllogism:

1. An actual infinite cannot exist.
2. An infinite temporal regress of events is an actual infinite.
3. Therefore, an infinite temporal regress of events cannot exist.[10]

This is given as part of the Kalâm argument and is supposed to thereby prove that the universe had to have a beginning. It doesn't, of course, because it is based on logic alone with no reference to empirical data.

Craig and Sinclair do not give scientists sufficient credit. You have to take a lot of mathematics to get a doctorate in physics, and we know what we are talking about. When we use "infinity" in physics, we simply mean "a very big number," not the abstraction Craig refers to as an "actual infinite."

In the nineteenth century, mathematician Georg Cantor placed the notion of infinity on a firm logical basis. He defined an infinite set as one whose elements can be put in one-to-one relation with a proper subset of itself. For example, the set of integers is an infinite set. It is limitless, but it contains no "final" number we identify as infinity.

Now, time is just what you measure on a clock. You count

the ticks to measure time, and if you keep on counting you will never reach infinity. You will just keep getting bigger and bigger numbers.

And you can count backward, too, and never reach the "beginning of time."

An infinite set can be small and bounded. For example, a second is a short time. But the number of moments, the number of distinct times, in one second is limitless. This is the kind of infinity that Hilbert was referring to as "an idea" that is not found in reality. We can't indefinitely measure finer and finer intervals of time, but it's a useful idealization.

Craig and Sinclair are invoking another kind of infinity: an infinity of measured extent. Their theorem above is correct. But it is irrelevant. While scientists are often sloppy in their use of the word *infinity*, they are not claiming that the universe is infinite in either space or time. There are saying it is limitless. Can

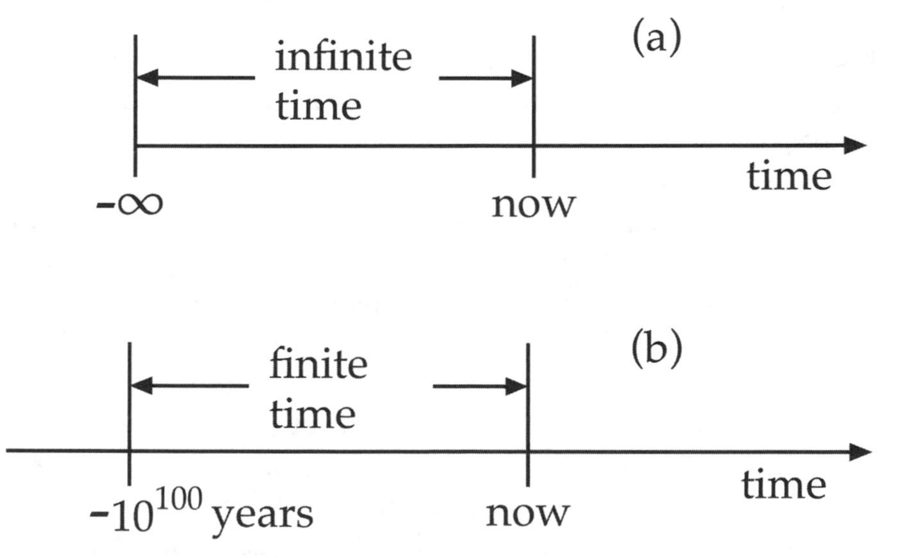

Fig. 6.1. (a) The eternal universe that Craig and Sinclair wrongly attribute to cosmology. Since it started an infinite time ago, it would have taken an infinite time to reach the present. (b) As viewed by cosmologists, there is no such thing as infinite time. The eternal universe had no beginning, not one an infinite time ago. No matter how many years you go into the past, it took a finite time to reach the present.

things stretch back arbitrarily far in time? Sure they can: yesterday, day before yesterday, one year ago, two years ago, a billion-billion years ago. The interval of time from a billion-billion years ago to now is not infinite, it is a denumerable billion-billion years.

Saying the universe is eternal simply is saying that it has no beginning or end, not that it had a beginning an infinite time ago (see fig. 6.1). Craig and Sinclair have not succeeded at this point—in proving the universe had a beginning.

6.2. THE MISSING SINGULARITY

For three decades now, until recently when he debated me on the campus of Oregon State University on March 1, 2010, Craig has used the argument that our universe had to begin as a *singularity*, an infinitesimal point of infinite density. Craig claims that time itself must have begun at that point, although I do not see why, since time is a human invention and we can start a clock at any time we want. Yet in virtually every book written by theists who bring up cosmology, we read this assertion. For example, Christian apologist and evangelist Ravi K. Zacharias states,

> Big Bang cosmology, along with Einstein's theory of relativity, implies that there is indeed an "in the beginning." All the data indicates a universe that is exploding from a point of infinite density.[11]

In his book, *What's So Great about Christianity?* Dinesh D'Souza writes,

> In a stunning confirmation of the book of Genesis, modern scientists have discovered that the universe was created in a primordial explosion of energy and light. Not only did the universe have a beginning *in* space and time, but the origin of the universe was also a beginning *for* space and time.[12]

However, the biblical story of creation bears no resemblance whatsoever to the big bang as described by modern cosmology. Genesis describes a creation taking place in six days a few thousand years ago. According to Genesis, Earth was created before the sun, the moon, and the stars within a fixed firmament. In contrast, scientific cosmology describes a universe tunneling out of chaos 13.7 billion years ago and an expanding universe, certainly not a "firmament," in which the solar system appeared 4.6 billion years ago, followed by the formation of the sun and Earth in the next 100 million years or so. In Genesis, humans appear shortly after the creation of Earth, on the sixth day. In fact, modern humans originated in Africa about 200,000 years ago. If humanity is so special in God's eye, doesn't it make you wonder why he waited 13.6998 billion years before creating us?

Now, maybe these are just details and the important point is that the Bible speaks of a creation in time. Well, just about every culture in human history has had some kind of myth of creation in time, so it is disingenuous for Christians to claim some unique insight in Genesis, especially since the insight is so grossly wrong in details!

D'Souza nonetheless insists on the theological significance of the big bang, referring to the teachings of St. Augustine (d. 430):

> There was no time before the creation, Augustine wrote, because the creation of the universe involved the creation of time itself. Modern physics has confirmed Augustine and the ancient understanding of the Jews and Christians.[13]

As I have noted, a beginning is necessary if you are to believe the universe was created. Of course, the universe may have had an uncreated beginning or a deistic creation, but if it had no beginning at all, no creator, a lot of theology has to be rewritten.

As we see, the argument for the big bang necessarily being the beginning of time is based on the notion that it originated in a "singularity," an infinitesimal point of infinite mass. Here's how Craig framed this argument in his debate with Pigliucci:[14]

> This conclusion [that the universe had a beginning] has been confirmed by remarkable discoveries in astronomy and astrophysics. The astrophysical evidence indicates that the universe began to exist in a great explosion called the "Big Bang" 15 [*sic*] billion years ago. Physical space and time were created in that event, as well as all the matter and energy in the universe. Therefore, as Cambridge astronomer Fred Hoyle points out, the Big Bang theory requires the creation of the universe from nothing. This is because, as you go back in time, you reach a point at which, in Hoyle's words, the universe was "shrunk down to nothing at all."[15] Thus, what the Big Bang model requires is that the universe began to exist and was created out of nothing.[16]

Although this idea has been discussed for years in the literature, Craig and other theologians generally have referred to a mathematical proof by Cambridge cosmologist Stephen Hawking and Oxford mathematician Roger Penrose published in 1970.[17] They showed that Einstein's theory of general relativity implied that the universe began with a singularity. Theologians then drew the conclusion that space and time themselves must have begun at that point and thus there was no time before the big bang.

The original Hawking-Penrose model is illustrated in figure 6.2. Shown is a space-time diagram with the vertical axis labeled "time" and two of the three spatial dimensions indicated on a perpendicular plane labeled "space." Of course, the universe has three spatial dimensions along with a fourth dimension of time, but this is difficult to illustrate on two-dimensional paper. So, while the universe is actually a three-dimensional sphere, its projection is shown in the figure as a circle perpendicular to the time

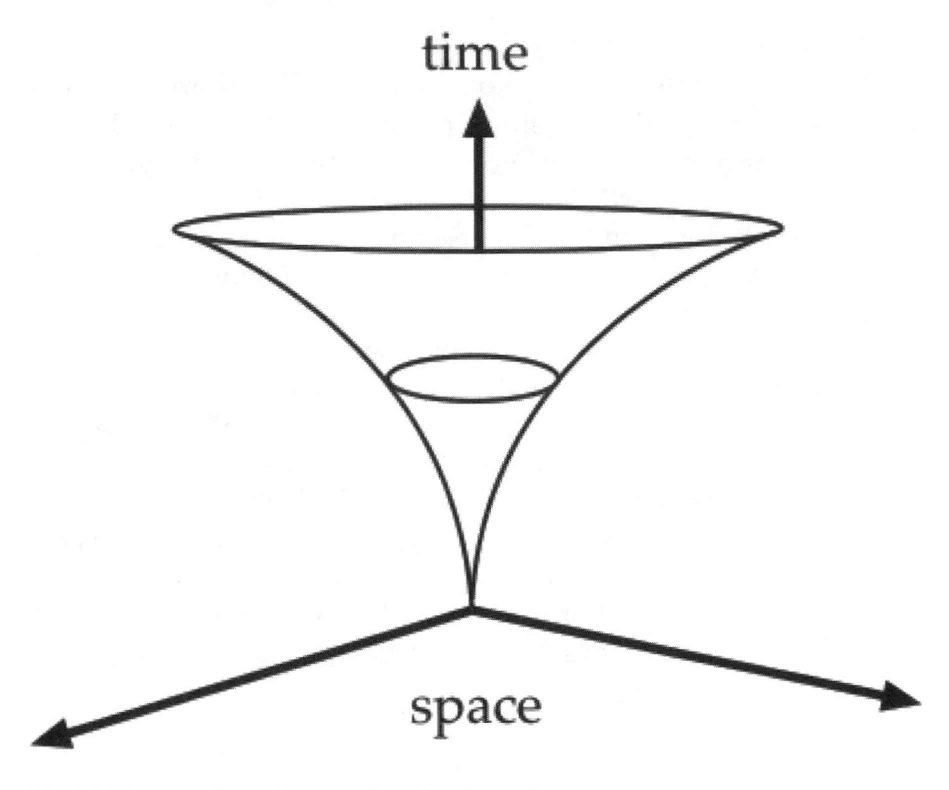

Fig. 6.2. A space-time diagram showing the early universe expanding from a point (singularity).

axis. The cone sweeps out these circles as time progresses. Only the early "inflationary" period in which the universe underwent exponential inflation is shown. (See chapter 5 for a discussion of inflation.)

At the origin, time equals zero, the cone contracts to a point. This is the singularity predicted by Hawking and Penrose.

Craig informs us, "Four of the world's most prominent astronomers described the beginning of the universe in these words":[18]

> The universe began from a state of infinite density. . . . Space and time were created in that event and so was all the matter in the universe. It is not meaningful to ask what happened before the Big Bang; it is like asking what is north of the North

> Pole. Similarly, it is not sensible to ask where the Big Bang took place. The point-universe was not an object isolated in space; it was the entire universe, and so the answer can only be that the Big Bang happened everywhere.[19]

It is to be noted that this quotation along with most of those Craig has used in his debates are over three decades old, although I still find it hard to believe that in 1976 one could find "four prominent astronomers" who had not thought of the obvious problem with the Hawking-Penrose prediction of a singularity: It does not occur when quantum mechanics is taken into account. General relativity breaks down before you reach zero distance and zero time.

Over twenty years ago, Hawking and Penrose agreed that a singularity did not, in fact, occur, and that is now the consensus of the working cosmological community. Hawking and Penrose's original calculation was not wrong as far as it followed from the assumptions of general relativity. But those assumptions had not taken into account quantum mechanics. General relativity, as successful as it has been in describing gravity, is not a quantum theory and does not apply when distances and times become very small, on the order of what is called the *Planck scale*. The *Planck length* is the smallest measurable distance, 1.616×10^{-35} meter. The *Planck time* is the smallest measurable time, 6.391×10^{-44} second. The universe was never an infinitesimal point in space-time.

In his blockbuster bestseller, *A Brief History of Time*, which came out in 1988, Hawking says: "There was in fact no singularity at the beginning of the universe."[20] Here is what he said in full:

> The final result was a joint paper by Penrose and myself in 1970, which at last proved that there must have been a big bang singularity provided only that general relativity is correct and the universe contains as much matter as we observe.[21]

Hawking continues:

> So in the end our work became generally accepted and nowadays nearly everyone assumes that the universe started with a big bang singularity. It is perhaps ironic that, having changed my mind, I am now trying to convince other physicists that there was in fact no singularity at the beginning of the universe—as we shall see later, it can disappear once quantum effects are taken into account.[22]

In his 2007 book *What's So Great about Christianity*,[23] Dinesh D'Souza lifted out of context the first part of Hawking's statement "that there must have been a big bang singularity." D'Souza then gave it exactly the opposite meaning to Hawking's intent. That intent becomes crystal clear upon reading on a few more lines: "I am now trying to convince other physicists that there was in fact no singularity at the beginning of the universe—as we shall see later, it can disappear once quantum effects are taken into account."[24]

So, instead of terminating at an infinitesimal point as in figure 6.2, the space-time cone that represents the early universe never reaches zero time, as shown in figure 6.3. That is, our universe begins with a finite size. Can we say anything about what went on "before"?

Although Craig continues to use the singularity argument in his debates with others, he specifically avoided it in his March 1, 2010, debate with me in Corvallis, Oregon. Instead, he brought up a new cosmological argument for the universe having a beginning. I will quote him exactly from his paper *Five Arguments for God*.[25]

> In 2003 Long Island University mathematician Arvind Borde, MIT physicist Alan Guth, and Tufts University cosmologist Alexander Vilenkin proved that any universe that is, on average, in a state of cosmic expansion cannot be eternal in the past but must have an absolute beginning. Their proof

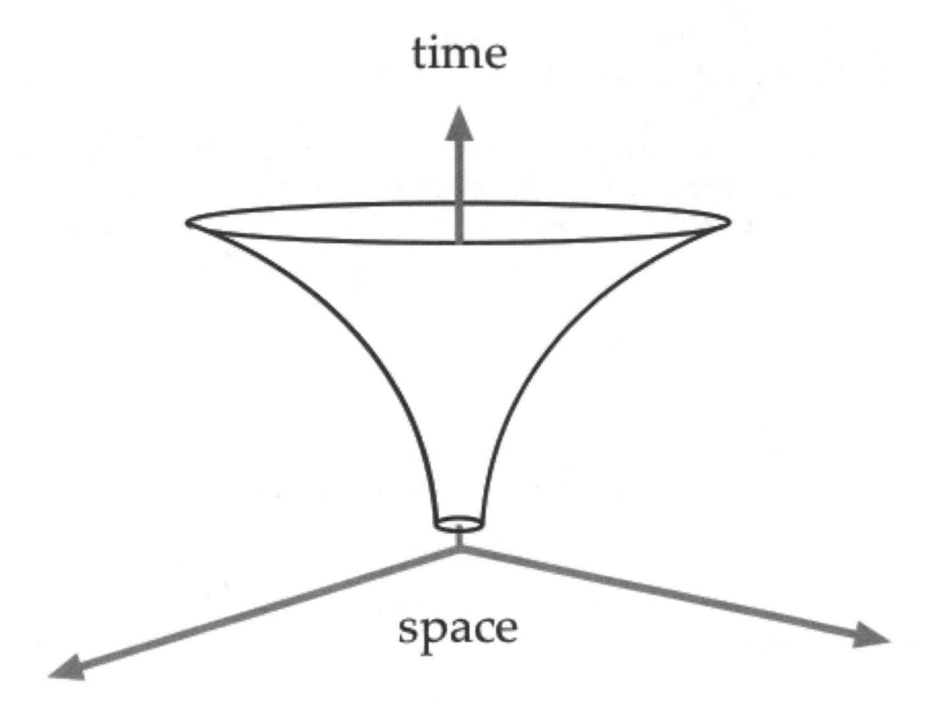

Fig. 6.3. A space-time diagram showing the expansion of the early universe taking into account quantum mechanics. There is no singularity. A universe of finite size begins at the Planck time.

> holds regardless of the physical description of the very early universe, which still eludes scientists, and applies even to any wider multiverse of which our universe might be thought to be a part.[26]

The conclusion that Borde and collaborators had proved that the universe had to have a beginning was disputed the same year by University of California–Santa Cruz physicist Anthony Aguirre and Cambridge astronomer Steven Gratton in a paper that Craig ignores.[27] Being good scholars, Borde et al. refer to Aguirre and Gratton in their own paper.

I contacted Aguirre and Vilenkin, the latter whom I have known professionally for many years. I greatly admire the work of each, which will be referred to often on these pages. I first asked Vilenkin if Craig's statement is accurate. Vilenkin replied:

> I would say this is basically correct, except the words "absolute beginning" do raise some red flags. The theorem says that if the universe is everywhere expanding (on average), then the histories of most particles cannot be extended to the infinite past. In other words, if we follow the trajectory of some particle to the past, we inevitably come to a point where the assumption of the theorem breaks down—that is, where the universe is no longer expanding. This is true for all particles, except perhaps a set of measure zero. In other words, there may be some (infinitely rare) particles whose histories are infinitely long.[28]

I sent this to Aguirre, who commented that the "infinitely rare" particles have worldlines [trajectories in space-time] that extend indefinitely into "the past," and can prevent there being a "time" at which the universe is not expanding/inflating. The fact that they are infinitely rare does not make them unimportant, because they nonetheless thread an infinite physical volume.[29]

I then asked Vilenkin, "Does your theorem prove that the universe must have had a beginning?" He immediately replied,

> No. But it proves that the expansion of the universe must have had a beginning. You can evade the theorem by postulating that the universe was contracting prior to some time.[30]

Vilenkin further explained,

> For example, Anthony in his work with Gratton, and Carroll and Chen,[31] proposed that the universe could be contracting before it started expanding. The boundary then corresponds to the moment (that Anthony referred to as $t = 0$) between the contraction and expansion phases, when the universe was

momentarily static. They postulated in addition that the arrow of time in the contracting part of space-time runs in the opposite way, so that entropy grows in both time directions from $t = 0$.

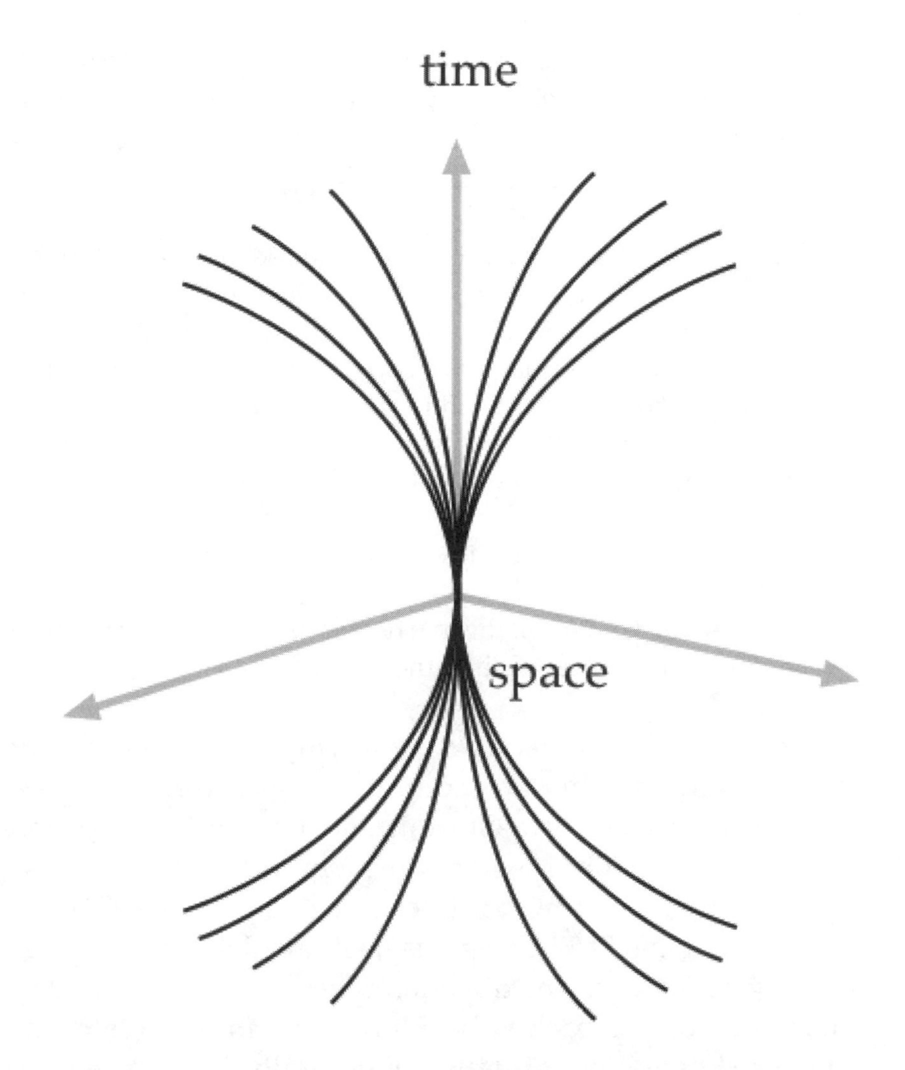

Fig. 6.4. Worldlines of particles are seen emerging from the origin during inflation. They can be extended in the negative time direction, thereby allowing for an eternal universe.

The problem and its solution are illustrated in figure 6.4. Worldlines of particles are seen emerging from the origin as part of inflation. Borde et al. proved they all had to come from a point, which then was interpreted as the beginning of the universe. Aguirre et al. showed that they can continue through the origin to the negative side of the time axis, allowing for an eternal universe.

I also checked with Caltech cosmologist Sean Carroll, whose recent book *From Eternity to Here* provides an excellent discussion of many of the problems associated with early universe cosmology.[32] Here was his response:

> I think my answer would be fairly concise: no result derived on the basis of classical spacetime can be used to derive anything truly fundamental, since classical general relativity isn't right. You need to quantize gravity. The BGV [Borde, Guth, Vilenkin] singularity theorem is certainly interesting and important, because it helps us understand where classical GR breaks down, but it doesn't help us decide what to do when it breaks down. Surely there's no need to throw up our hands and declare that this puzzle can't be resolved within a materialist framework.
>
> Invoking God to fill this particular gap is just as premature and unwarranted as all the other gaps.[33]

The next day, Vilenkin wrote to emphasize there are problems with the stability of a contracting universe, which is why the two teams, Aguirre-Gratton and Carroll-Chen, proposed that the arrow of time switch directions at $t = 0$.

Actually, the arrow of time is defined by the increase in entropy, so the "switching" is just by definition, since the entropy increases in the direction of expansion.[34] I will discuss this notion of two back-to-back universes in detail later in this chapter, showing how it falls out naturally from the equations of cosmology.

In their long article in *The Blackwell Companion to Natural*

Theology, Craig and Sinclair continue to press many of the arguments I have refuted here. They still refer to conclusions drawn from general relativity that simply do not apply in the quantum regime around $t = 0$.

They specifically object to the solution to the BGV theorem proposed by Aguirre-Gratton and Carroll-Chen described above. They argue:

> It is possible, then, to evade the BGV theorem through a gross deconstruction of the notion of time. Suppose one asserts that in the past contracting phase the direction of time is reversed. Time then flows in both directions away from the *singularity* [my emphasis]. Is this reasonable? We suggest *not*, for the Aguirre-Gratton scenario denies the evolutionary continuity of the universe which is topologically prior to t and our universe. *The other side of the de Sitter space is not in our past*. For the moments of that time are not earlier than t or any of the moments later than t in our universe. There is no connection or temporal relation whatsoever of our universe to that other reality. Efforts to deconstruct time thus fundamentally reject the evolutionary paradigm.[35]

I sent this on to Aguirre for comment. Here was his response:

> First, there is no singularity—that is the whole point. Second, he simply asserts that "the evolutionary continuity of the universe [[is]] topologically prior to t and our universe." And I'm not really sure what this means. I would agree in some sense that "there is no connection or temporal relation whatsoever of our universe to that other reality." But I'm not sure why this matters. In the Aguirre-Gratton type model, the idea is to generate a steady state. If you do so, you can use the BGV theorem to show that there is a boundary to it. We have specified what would have to be on that boundary to be consistent with the steady state, and argued that this can be nonsingular, and also defines a similar region on the other side of that

> boundary (which in fact might be identifiable with our side, topologically). There is no "initial time," nor any "time" at which the arrow of time changes. In the natural definition of time in this model, there is no earliest time, and all times are (statistically) equivalent.[36]

Let me give my personal reaction. If you look at the scenario I presented earlier in chapter 5, you will see that the Friedmann equations for an empty universe predict inflation on both sides of t-axis around and away from $t = 0$. As Aguirre notes, there is no singularity. Instead, there is a region of total chaos around $t = 0$. So there is no problem of continuity and an abrupt switch of the arrow of time. The definition of the arrow of time is arbitrary.

6.3. QUANTUM TUNNELING

Continuing to quote from his debate with Pigliucci, which we recall still relied on the long-refuted singularity argument, Craig says,

> Now this tends to be very awkward for the atheist. For as Anthony Kenny of Oxford University urges, "A proponent of the big bang theory, at least if he is an atheist, must believe that the . . . universe came from nothing and by nothing."[37] But surely that doesn't make sense! Out of nothing, nothing comes. So why does the universe exist instead of just nothing? Where did it come from? There must have been a cause which brought the universe into being.[38]

In the early 1980s, several published papers showing how our universe could have arisen naturally by a well-established process known as *quantum tunneling* appeared.[39] Many more proposals have appeared since then, but allow me to concentrate on these older schemes, since they are easier to understand in terms of established physics. Quantum tunneling is important to understand for our purposes, so let us first take a moment to understand that phenomenon.

In physics, the total mechanical energy of a particle is equal to the sum of its kinetic energy and potential energy: $E = K + V$. (To get the total energy we must add the rest energy.) Consider a particle of positive mechanical energy E and zero potential energy, $V = 0$. Its kinetic energy will be $K = E$. Suppose this particle moves toward a potential energy barrier of height $V > E$, as shown in figure 6.5. According to classical mechanics, the particle cannot surmount the barrier because it does not have sufficient energy.

In quantum tunneling, a particle is able to pass through the barrier. In figure 6.5, a particle is shown coming in from the left toward the barrier. The dashed line gives the energy of the particle, which we see is below the height of the barrier. Also plotted on the same graph is the wave function, which, to the left of the barrier, is a sinusoidal wave.

Inside the barrier, the kinetic energy of the particle, $K = E - V$, is negative and therefore the square of the momentum of the particle, $p^2 = 2mK$, is also negative. The momentum $p = \sqrt{p^2}$ is mathematically an *imaginary number*, which is a number containing a factor of $\sqrt{-1}$. When you square $\sqrt{-1}$, you get -1. The familiar *real number* has a square that is always positive. The set of a real number and imaginary number is called a *complex number*, often written $c = a + ib$, where a is the *real part*, b is the *imaginary part*, and i is $\sqrt{-1}$. It can also be written $A\exp(i\phi)$, where $A = \sqrt{a^2 + b^2}$ is the *amplitude* and ϕ is the *phase*. The particle is thus termed "unphysical" because it does not have the normal mathematics of a measurable particle; but we still can describe it mathematically.

Complex numbers have amazing properties and are widely used in science and engineering. The wave function is, in general, a complex number, which presents no problem since it is an abstract quantity that does not correspond to any observable quantity within quantum theory. The wavelike interference effects in quantum mechanics are described in terms of complex numbers.

The time-independent Schrödinger equation that is used to calculate the wave function has a nonzero solution inside the barrier. The wave function is no longer a sinusoid but has the form of an exponential. If the barrier is not too thick, the wave function will not have decayed to zero when it reaches the right edge. Upon passing though, it resumes the oscillatory form associated with a measurable particle since the kinetic energy is once again positive and the momentum is again a real number. Thus the particle has "tunneled" through the barrier.

Quantum tunneling has been known for over fifty years and is utilized in the *scanning tunneling microscope,* which is an instrument capable of producing a digital image of individual atoms. Indeed, without quantum tunneling we would not be here. By means of quantum tunneling the protons in the sun are

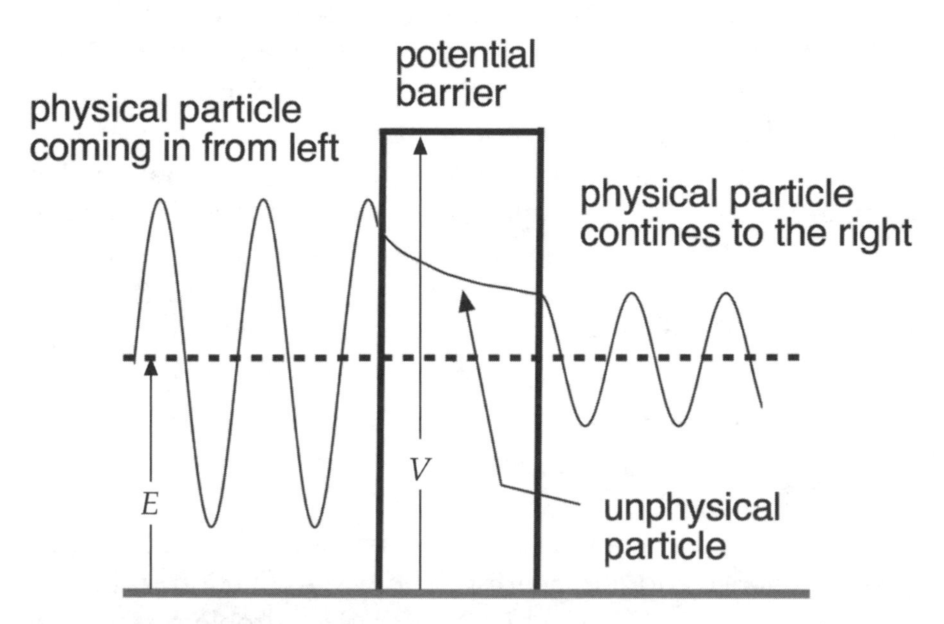

Fig. 6.5. Quantum tunneling through a barrier. Inside the barrier the particle is "unphysical" but becomes physical again when it emerges from the other side of the barrier.

able to penetrate the barrier produced by their mutual electrical repulsion and merge with neutrons to produce helium nuclei by the process called *nuclear fusion*. This is the source of energy in the sun and all the stars in the sky.

6.4. IMAGINARY TIME

I mentioned above how in a 2007 book Dinesh D'Souza lifted a quotation by Stephen Hawking out of context to try to make it seem that Hawking still supported the notion of a singularity at the beginning of the universe.[40] In a follow-up 2009 book, *Life after Death: The Evidence*, D'Souza attempts to demean what is, in fact, the trivial explanation for the absence of the singularity that I gave above. There are no space-time singularities in quantum mechanics because there are no space-time points. D'Souza tells us that scientists are "working hard to come up with an explanation for the big bang that avoids having to posit a creator."[41] I guess someone who worked in politics for so many years finds it hard to believe that there actually are people in the world who are motivated by the truth rather than winning points over the opposition. He certainly does not understand how science works, how a scientist must go wherever the data lead regardless of personal wishes. Of course, there are examples of scientific fraud where a scientist breaks this rule. But they are rare because of the self-correcting nature of science.

This is to be contrasted to Christian apologists, who ignore the data that do not support their beliefs and work hard at mining scientific quotations that seem to support them. D'Souza goes on,

> Taking up the challenge, Stephen Hawking has proposed a scenario in which the universe could have come into existence without an original singularity. Hawking's proposal involves something called "imaginary time," a mathematical concept

referring to the square root of a negative number. Nothing in the world is known to operate in imaginary time.[42]

We saw above that a particle inside a barrier has imaginary momentum but still can be described by a mathematically proper wave function. Furthermore, and more important, the prediction of quantum theory that the particle can tunnel though the barrier to the other side, a process not allowed in classical physics, has been thoroughly confirmed by observations for over fifty years.

While it is true that we cannot measure the particle's momentum inside the barrier, we can still treat it mathematically. This demonstrates that physicists are not limited to speaking only about that which they can directly observe and measure. The mathematics that describes observable events can also be used to describe unobservable events and provide useful predictions of other events that are observable. Following convention, I will call a region with observable events "physical" and a region with unobservable events "unphysical."

Let us assume a universe with no matter or radiation, just the cosmological constant. And, for a reason I will later explain, let us assume a closed universe, $k = 1$. The Friedmann equations then give as a solution for the cosmic scale factor (or simply the radius of the universe in the semi-Newtonian model)

$$a(t) = a_o \cosh\left(\frac{t}{a_o}\right) \tag{6.1}$$

where $a_o = \sqrt{3/\Lambda}$. (I will not consider the case where Λ is negative.) We can write this

$$a(t) = \frac{a_o}{2}\left[\exp(\frac{t}{a_0}) + \exp(-\frac{t}{a_0})\right] \tag{6.2}$$

When $t > 0$, the second term in (6.2) goes rapidly to zero, and we have the inflationary expansion of the early universe described in chapter 5 for $k = 0$. Note that nothing excludes the region $t < 0$, where the first term goes rapidly to zero and the second term gives exponential inflation in the negative time direction.

Now, we still have not explained imaginary time. The Friedmann equation allows for an unphysical range of the cosmic scale factor a in which the time variable is imaginary. For that solution the scale factor can take on values less than the value a_0 the scale factor has at $t = 0$. Those solutions have imaginary time $\tau = it$.

The solution (6.1) tells us that $a = a_0$ at $t = 0$ and $a > a_0$ for $t > 0$. However, there is another solution. For $a < a_0$ let $\tau = it$. This is the infamous *imaginary time*. Then (6.1) becomes

$$a(\tau) = a_0 \cos\left(\frac{\tau}{a_0}\right) \tag{6.3}$$

which applies when $a < a_0$.

Now, there is an interesting thing about imaginary time. It makes space-time Euclidean. Recall from chapter 3 that the proper distance in space-time is $s^2 = -(ict)^2 + x^2 + y^2 + z^2$. If $\tau = it$, $s^2 = c^2\tau^2 + x^2 + y^2 + z^2$ and we have a Euclidean space-time in which the proper distance s is just the radius of a four-dimensional sphere. For simplicity, let us illustrate a three-dimensional Euclidean space-time with two spatial axes and one time axis, suppressing the third spatial axis for clarity. Then, as seen in figure 6.6, the unphysical early universe is described as a three-dimensional sphere with space a two-dimensional circle of radius a as before. Time now is where that circle intersects the circle of longitude. The third dimension of space is again suppressed for clarity.

In the model proposed by Hawking and University of California–Santa Barbara physicist James Hartle, which will be described in detail below, we picture the universe contracting down toward the South Pole. The circle that represents space decreases until it becomes a point at the South Pole (we are not using quantum mechanics here, just general relativity). That point is not a singularity, however, and does not represent any

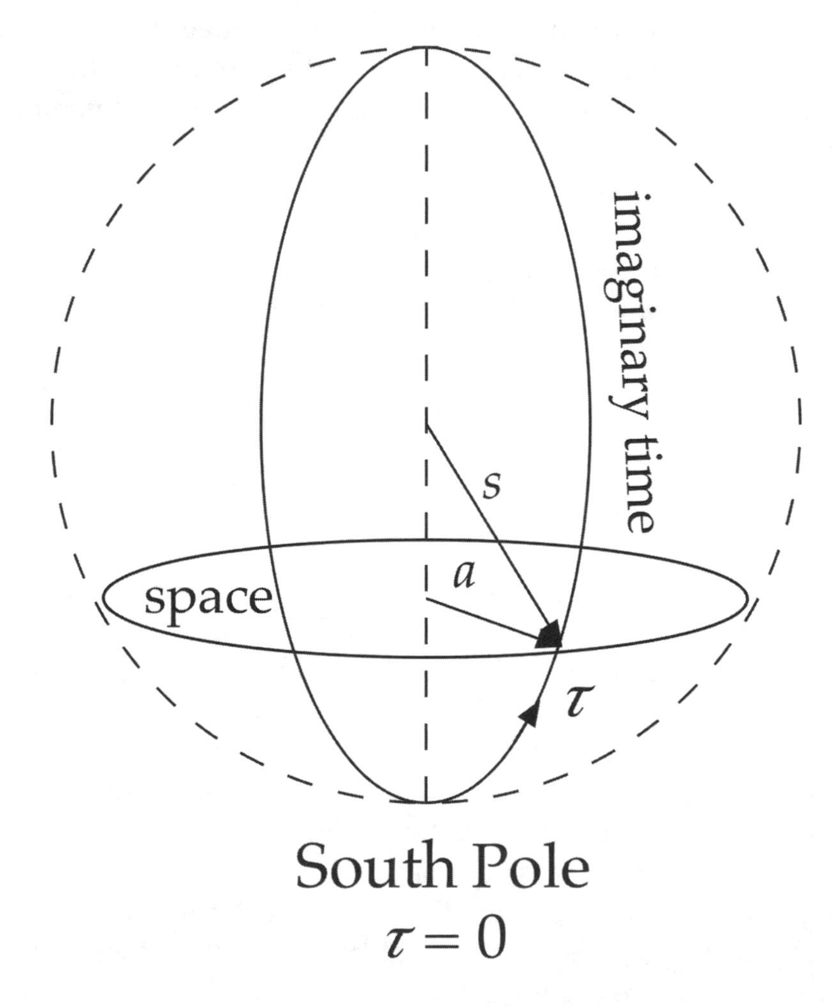

Fig. 6.6. Euclidean space-time with imaginary time.

special moment of time. It is just a space-time point like any other on the sphere. The universe passes through the South Pole and time continues up along the longitude circle in the direction of the arrow shown in the figure. When it reaches the equator, the universe looks like it will start contracting and end up as a spatial point at the North Pole and then start expanding again. However, before this happens, the radius increases to the point where we enter the physical region $a > a_0$ and the universe pops into existence from what, for all practical purposes, appears as "nothing."

Let us stay in the unphysical region a little longer and provide an alternative view that is completely consistent with the same equations. We label the South Pole $\tau = 0$ and then have two universes going in opposite time directions.

Now, so far this has been a purely classical analysis. To give the unphysical region meaning, we have to bring in quantum mechanics.

6.5. A SCENARIO FOR A NATURAL ORIGIN OF THE UNIVERSE

In the case of the quantum tunneling models of the origin of our universe, a *wave function of the universe* can be calculated. This is a rather esoteric concept, but it is really no different in principle than calculating the wave function for a typical quantum system, such as a hydrogen atom. In a very useful 1994 review paper called "Quantum Cosmology for Pedestrians," Skidmore College physicist David Atkatz derived the wave functions for two of the models using comparatively simple, undergraduate-level physics and mathematics.[43]

In an earlier paper, Atkatz and Heinz Pagels had showed that quantum tunneling was only possible with the cosmological curvature parameter $k = 1$.[44] As we saw above with equation (6.2), inflation is still predicted in this case. That is, inflation

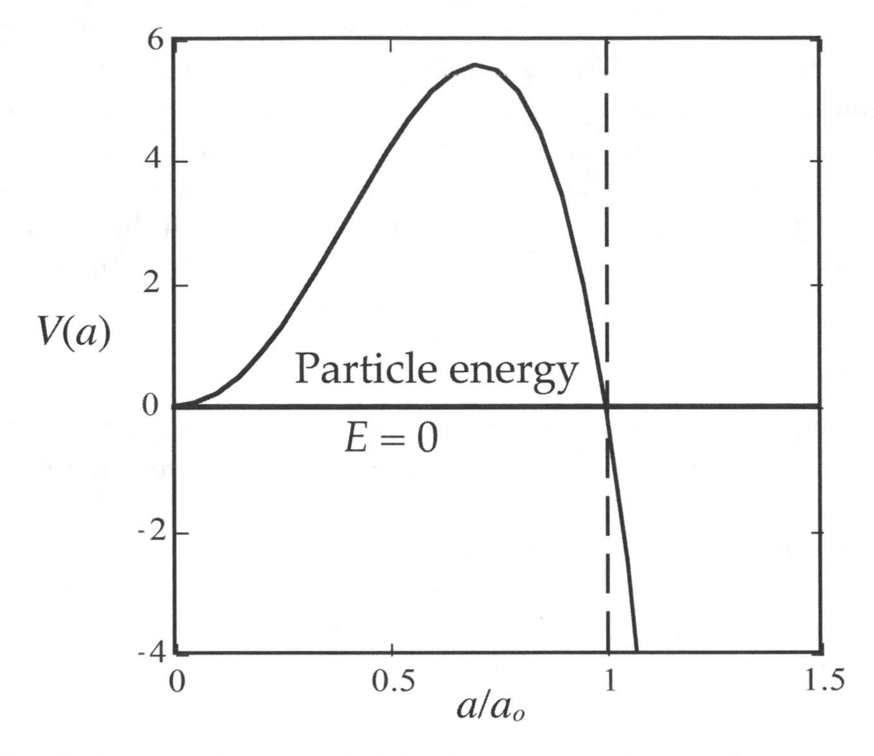

Fig. 6.7. The potential energy in the Schrödinger equation that gives the wave function of the universe. The region $a < a_0$ is unphysical. $V(a)$ presents a barrier to a particle of zero energy coming in from the right.

does not require $k = 0$, as is commonly believed. The universe can start out with some curvature and still flatten out considerably, like a large expanding balloon, after many orders of magnitude of exponential inflation.

Starting with the classical Friedmann equations of cosmology that follow from general relativity and setting $k = 1$, we follow the standard *quantization* procedure that takes you from an equation in classical mechanics to the corresponding equation in quantum mechanics. A homogeneous, isotropic universe is assumed so the only variable is the cosmic scale factor $a(t)$ that we can roughly think of as the radius of the universe. We obtain Schrödinger's nonrelativistic, time-independent equation for a particle of mass equal to half the Planck mass and energy zero whose potential energy is given in figure 6.7.

$$\left[\frac{d^2}{da^2} - \left(\frac{3\pi}{2G}\right)^2 a^2\left(1 - \frac{a^2}{a_o^2}\right)\right]\psi = 0 \qquad (6.4)$$

where $a_o^2 = \dfrac{3}{8\pi\rho G}$.

The energy density of the universe is given by ρ and $\hbar = c = 1$.

The state ψ is called the "wave function of the universe." Equation (6.4) is a special case of the Wheeler-Dewitt equation.[45]

Note that the plot in figure 6.7 includes the unphysical region $a < a_o$ where time is imaginary. In that region, the potential energy forms a barrier that prevents the penetration of a classical particle but can be tunneled through quantum mechanically.

Atkatz reports two different explicit solutions for the wave function. One, provided by Vilenkin in 1986, assumes the universe started at $t = 0$ with the wave function zero and just an outgoing wave function.[46] The other, provided by Hartle and Hawking in 1983, assumes both incoming and outgoing waves.[47] They called this the *no boundary condition model*. Recall our discussion above that described the Hartle-Hawking scenario in terms of imaginary time. Figure 6.8 shows the wave function for the model of Hartle and Hawking. Notice it depends only on a. The wave function of the universe so calculated does not depend explicitly on time.

Now, picture the universe collapsing from a larger value of a. The zero-energy particle that represents the universe in the Schrödinger equation will bump into that barrier at $a = a_0$.

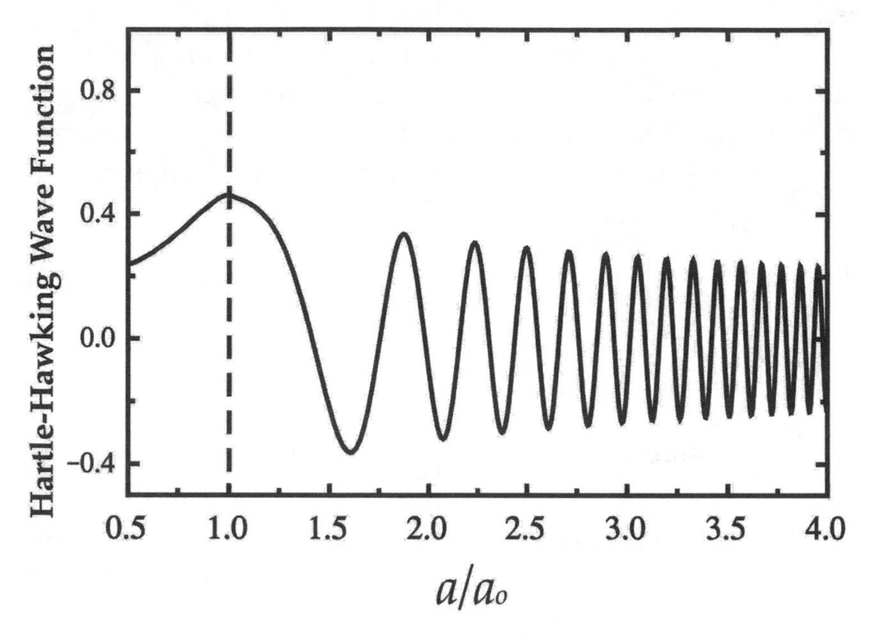

Fig. 6.8. The Hartle-Hawking wave function of the universe. For $a > a_0$ the particle representing the universe is physical, with a real momentum equal to zero. For $a > a_0$ it is unphysical with an imaginary momentum.

Assuming the conventional interpretation of the wave function as used in quantum mechanics, the square of the magnitude of the wave function gives the probability that a universe in an ensemble of universes, if you can imagine such a thing, will be found at a particular radius, per unit radius.

We illustrate the tunneling scenario in figure 6.9. Our universe appears by quantum tunneling from the unphysical region.

6.6. THE BIVERSE

As we saw above, Hartle and Hawking developed what they called the no boundary condition model because it did not assume the universe began at zero time. They originally pictured the wave function as coming in from infinity, bouncing off

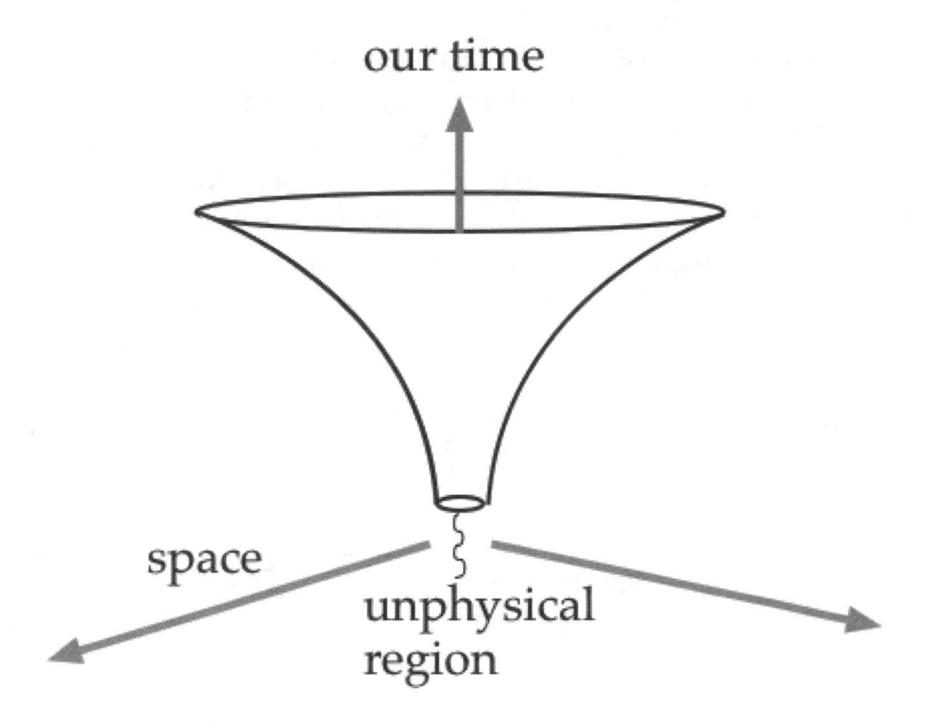

Fig. 6.9. Our universe appears by quantum tunneling from an unphysical region.

the barrier, and going back to infinity. This they described in terms of the picture in figure 6.6, where the point on the great circle representing imaginary time passes through the South Pole and continues to increase.

However, the Hartle-Hawking model has another interpretation that is mathematically equivalent, namely, one in which two universes start at $t = 0$ and proceed in opposite directions in time. Hawking mentions the possibility in *300 Years of Gravitation*, published in 1987.[48] I have promoted the notion in several popular books and articles going back to 2000.[49]

So the well-established equations of physics and cosmology allow for the existence of two mirror universes: ours that expands along the positive time axis and a prior universe that exists at negative times. From our point of view, our universe appears by quantum tunneling from the earlier universe. This is illustrated in figure 6.10. From the point of view of observers in the mirror universe, their universe quantum tunneled from ours.

Now, our sister universe is contracting from our point of view. However, the direction or "arrow" of time is defined in physics as the direction of increasing entropy in the universe, which is opposite to ours in the other universe. Thus in that universe we must reverse the arrow of time so it points opposite to ours.

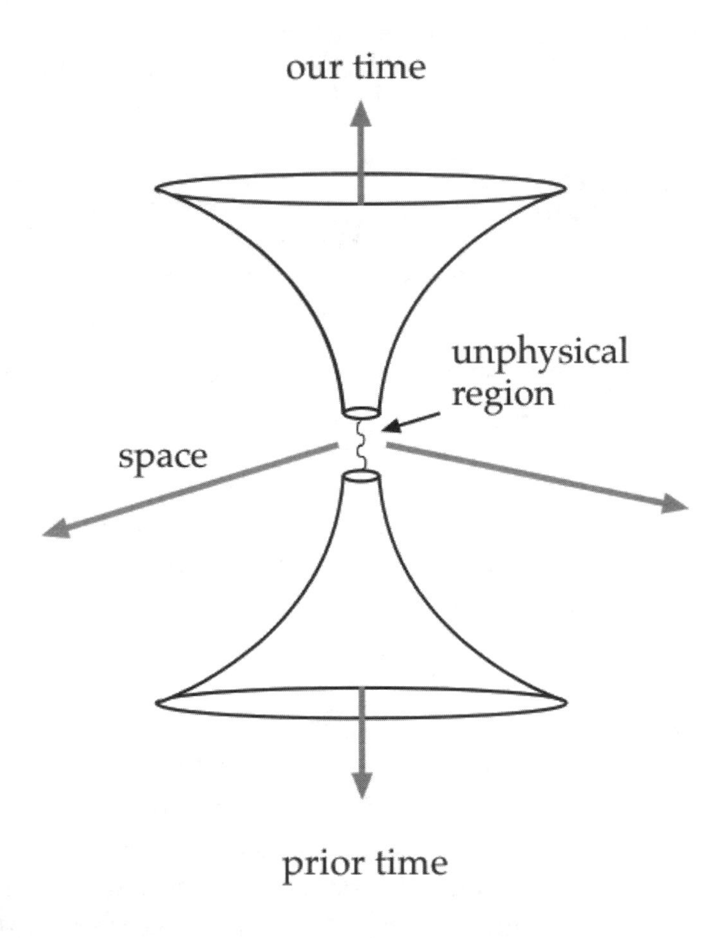

Fig. 6.10. Our universe appears by quantum tunneling from the prior one, from our point of view. However, time points in the opposite direction in our mirror universe, so both universes emerge from the same unphysical region.

The two opposite arrows of time solve a number of problems. Recall my extensive discussion earlier in this chapter on the BGV theorem, which basically says inflation had to have a beginning. This has been used by William Lane Craig to argue that the universe itself had to have a beginning. We saw that cosmologists I contacted, including Vilenkin, Carroll, and Aguirre, all of whom have published works on the subject, agreed that no such conclusion is warranted. We just have a universe prior to inflation. The biverse scenario, which I have described above, was also suggested by Carroll and his collaborator, Chen, and by Aguirre and his collaborator, Gratton. See the discussion by Carroll in his 2010 popular book *From Eternity to Here*.[50]

The other beauty of the biverse scenario is its symmetry. It is invariant under time-reversal (as well as particle-antiparticle conjugation *C* and parity *P*). It also explains why the universe has low entropy at $t = 0$. As we saw in chapter 5, the universes start (in both time directions), by definition, with lowest entropy, from the smallest operationally defined volume, a sphere of Planck dimensions. Such a sphere has the *minimum* entropy that a sphere of its size can have. It also has the *maximum* entropy it can have, so it is in total chaos. As the sphere expands, the maximum allowable entropy in each universe expands, as shown in figure 6.11. Recall from chapter 5 that order can form in these universes since the local entropy is always less than maximum.

Carroll, however, is not ready to buy into the idea of the biverse because it does not explain why the entropy of the universe was so low at t=0.[51] Frankly, I don't see the problem. The entropy is maximum at that point as well as at both ends of the time axis. It is lower at $t = 0$ because the definition of the arrow of time is the direction of higher entropy. In other words, its lower where it is by definition.

While I cannot prove that this is actually how our universe came to be, we can say that a complete scenario for the natural origin of the universe that is consistent with all known physics and cosmology can be written out mathematically. While each universe had a beginning, that beginning did not require a creator. These beginnings occurred by uncaused quantum tunneling from nothing. Or, if you can't grasp the notion of tunneling from nothing, just think of our universe as having tunneled from an earlier one, a process that is already well understood.

You will often read about the universe beginning as a "quantum fluctuation." This is usually accompanied by a handwaving reference to the fact that particle-antiparticle pairs can appear and disappear spontaneously in a vacuum.

Indeed, this process, called "vacuum polarization," is an important part of quantum electrodynamics (QED), which deals with electromagnetic interactions, such as those between

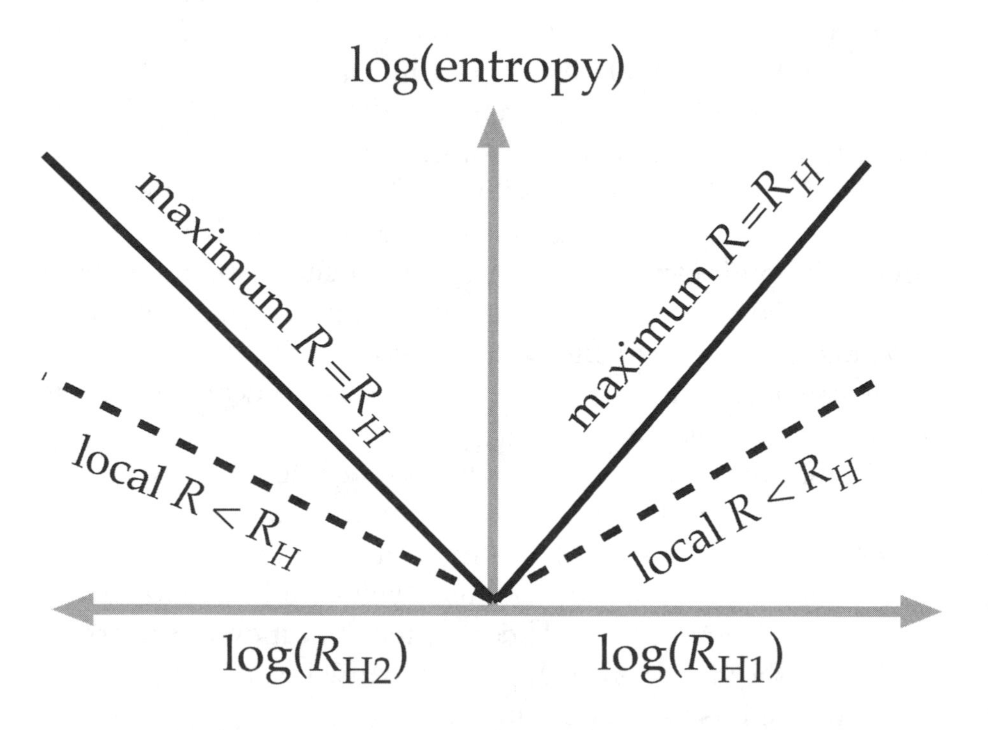

Fig. 6. 11. The maximum and actual entropy of the two universes in the biverse as both universes expand in opposite directions. While they start with maximum entropy, the expanding universes have increasingly less than maximum local entropy leaving room for order to form.

photons and electrons. An analogous process occurs in quantum chromodynamics (QCD), which deals with strong nuclear interactions, such as those between protons and neutrons (more fundamentally, between quarks and gluons).

However, this allusion is misleading. Vacuum polarization occurs in the presences of other bodies. If it occurs in the absence of any bodies, in a really empty vacuum, it can contribute nothing, since the process conserves energy exactly and it came from a state of zero energy.

On the other hand, quantum tunneling can be crudely understood as a quantum effect that results from the Heisenberg uncertainty principle: a temporary upward fluctuation in the particle's energy sends it over the barrier.

Since the scenario presented agrees with all known scientific principles, it cannot be said that some law of nature was necessarily violated in order to bring our universe into existence.

Nevertheless, I need to make it clear that neither the biverse nor a multiverse is required to demonstrate the fallacy of fine-tuning. At the same time, I reject the charge that it is not scientific to consider them even if they are not observable. They follow from models of physics cosmology that have been amply verified and, as I have demonstrated, the mathematics of these models can be used to extrapolate beyond observable domains to obtain a hint of what may be out there.

7.

Gravity Is Fiction

7.1. GRAVITY AND ELECTROMAGNETISM

Let us begin to look at the serious contenders for fine-tuning with the large number puzzle. This conundrum appears in virtually every book and article written on fine-tuning. Recall from chapter 1 that the large number puzzle was historically the first of the anthropic coincidences. In 1919, Hermann Weyl wondered why the ratio of the electromagnetic force to the gravitational force between an electron and a proton is such a huge number, $N_1 = 10^{39}$. He conjectured that any natural dimensionless number should not be too far from unity.

Let us calculate N_1. The static electrical force between two particles of electric charges q_1 and q_2 separated by a distance r is given by *Coulomb's law of electricity*, which says that the force between two point charges is proportional to the product of the charges and inversely proportional to the square of the distance between them. The force is repulsive for like charges and attractive for unlike charges. The gravitational force between two particles of masses m_1 and m_2 separated by a distance r is given by *Newton's law of gravity*, which says that the force between two point masses is proportional to the product of the masses and inversely proportional to the square of the distance between them. The force is always attractive, although, as we saw in chapter 5, Einstein's general theory of relativity, which is the advanced theory of gravity, allows for the possibility of repulsive gravity. This possibility was confirmed in 1998 by the observed acceleration of the expansion of the universe. But this does not concern us here.

The electric and gravitational forces according to Coulomb's and Newton's laws are both inverse square forces, so if one computes the ratio of the forces, the distances cancel. For the electron and the proton, the electric force is 10^{39} times stronger than the gravitational force.

The ratio of forces is

$$N_1 = \frac{F_E}{F_G} = \frac{k_E \frac{q_1 q_2}{r^2}}{G \frac{m_1 m_2}{r^2}} = \frac{k_E q_1 q_2}{G m_1 m_2} \qquad (7.1)$$

Here k_E is *Coulomb's constant* (8.987×10^9 Nm^2/C^2) and G is *Newton's constant* ($G = 6.674 \times 10^{-11}$ Nm^2/kg^2). Note that the r's cancel out so the result does not depend on the distance between the two particles.

For a proton and an electron, $q_1 = q_2 = e = 1.602 \text{x} 10^{-19}$ C, the unit electric charge, $m_1 = m_e = 9.109 \times 10^{-31}$ kg, and $m_2 = m_p = 1.673 \times 10^{-27}$ kg. Then,

$$N_1 = \frac{k_E e^2}{G m_e m_p} \qquad (7.2)$$

Plugging in the numbers, we get $N_1 = 10^{39}$, where an order of magnitude is sufficient for our purposes.

If the gravitational force between elementary particles were not much smaller than the electrical force, then the universe would collapse long before there was a chance for stars to form and life to evolve. So, it is argued, the value for N_1 above is highly improbable and must have been fine-tuned for life.

7.2. NOT UNIVERSAL

Note that N_1 is not a universal number; it depends on the charges and masses of the bodies you use in the calculation. For an electron and proton the ratio is the famous 10^{39} that I will continue to call N_1. But why assume these two particles in defining N_1? The proton is not even fundamental. It is made of quarks. For two electrons, the ratio is 10^{47}. If we have two unit-charged particles of equal mass 1.85×10^{-9} kilograms, $N_1 = 1$ and the two forces would be equal!

If we were to ask what mass is the most fundamental, it would be the *Planck mass*, which is formed from the fundamental constants and equals 2.18×10^{-8} kg. The gravitational force between two particles, each with the Planck mass and unit electric charge, is 137 times stronger than the electric force!

$$m_{Pl} = \sqrt{\frac{\hbar c}{G}} \qquad (7.3)$$

In most physics textbooks you will read that gravity is the weakest of all forces, many orders of magnitude weaker than electromagnetism. Christian philosopher Robin Collins makes the same assertion: "Compared with the total range of the strengths of the other forces in nature (which span a range of 10^{40}), it [the gravitational force] is very small, being one part in 10 thousand, billion, billion, billion."[1]

We see this is wrong. Recall that the gravitational force is fictional, like the centrifugal force. No one compares the value of centrifugal force with other forces. Its value depends on the circumstances. So it is with gravity. The gravitational force depends on the masses and charges of the particles. N_1 is only the ratio of the two forces for a system made of a proton and an electron, as in the hydrogen atom. It is not the relative strength of the gravitational and electrical forces in all cases. In fact, there is no universal way we can describe the strength of the gravitational force. The strength of the electromagnetic force[2] is measured by the dimensionless parameter α, called historically the *fine structure constant*, which in standard international units is $\alpha = k_E e^2 / \hbar c$. The value of α in our universe is currently 1/137 independent of units. In natural units, $\hbar = c = k_E = 1$, we simply have $\alpha = e^2$. In the standard model of elementary particles and forces, α actually varies with energy, but we will put off that discussion until later.

The gravitational constant G is not dimensionless and, as we have seen, is an arbitrary number, like $\hbar$ and c, that just sets the unit system. We can define a dimensionless parameter $\alpha_G = Gm^2 / \hbar c$ to represent the gravitational force strength, but that depends on some mass m. Conventionally, the proton mass is used, so for $m = m_p$, $\alpha_G = 5.01 \times 10^{39}$. However, as already noted, the proton is not even a fundamental particle but is composed of quarks. In short, the strength of gravity is an arbitrary number and is clearly not fine-tuned. It can be anything we want it to be.

This does not mean that the strength of gravity relative to the other forces is not important. It just depends numerically on how you define it. That definition does not change the ratio of the forces between two particles in any specific situation.

7.3. WHY ARE MASSES SO SMALL?

From the above discussion it is clear that the large number puzzle is not that N_1 is so large but that the masses of elementary particles are so small. Suppose they were of the order of the

Planck mass. Gravity would then be 137 times stronger than electromagnetism. In that case, there would be no long-lived structures, such as galaxies and stars, in which life could evolve. But elementary particles have such low masses that gravity is very weak compared to the other forces, and galaxies and stars do not collapse but have very long lives. Right now elementary particle masses are determined by experiment, but we can still give reasons why they are so low.

Let us inquire where mass comes from. In the current standard model, the fundamental particles are the quarks, leptons, and gauge bosons (recall table 4.1). The gauge bosons are the spin 1 particles that "mediate" the various interactions. For example, two quarks interact with each other by, crudely speaking, tossing gluons back and forth. The principle of *gauge invariance*, which was covered in chapter 4, demands that the masses of the gauge bosons be exactly zero.[3] This is true for the photon that mediates the electromagnetic force and the gluon that mediates the strong nuclear force. However, the W- and Z-bosons that mediate the weak nuclear force result from broken gauge symmetry and take on mass by a process called the *Higgs mechanism* (after one of its inventors, physicist Peter Higgs).

7.4. HOW TO GET MASS

In 1964, Higgs proposed that the universe is filled with massive spin zero particles now called *Higgs bosons*. We can crudely understand how this leads to mass by glancing back at figure 3.1. There we saw the zigzag path of a photon through a medium and how it results in an effective speed of light in a medium that is less than the vacuum speed *c*. Now, in the case of the medium being a field of Higgs bosons, the photon actually will not zigzag, since it does not interact with the Higgs boson in the standard model. However, the other particles in the standard model do interact and produce zigzag paths through the Higgs field.

Mass is the measure of inertia, that is, the sluggishness of a body. The more massive a body, the harder it is to get it moving, to stop it, or in general to change its velocity. The zigzagging particle is more sluggish in moving through a medium compared to a particle that moves in a straight line. That zigzagging might be at small angles so it is not as obvious as in the figure, but it can be sufficient to slow the particle down.

Besides the gauge bosons (photon and gluons excepted), electrons, muons, and tauons all pick up mass by the Higgs mechanism. Quarks must pick up some of their masses this way, but they obtain most of their masses by way of the strong interaction, as will be discussed below. The three types of neutrinos may gain very tiny masses by a special process known as the "seesaw" mechanism. All these masses are orders of magnitude less than the Planck mass, and no fine-tuning was necessary to make gravity much weaker than electromagnetism. This happened naturally and would have occurred for a wide range of mass values, which, after all, are just small corrections to their intrinsically zero masses.

As of this writing, the existence of the Higgs particle has still not been empirically confirmed or falsified, although the time is now approaching when we should have a definitive indication one way or the other as the Large Hadron Collider (LHC) goes into operation in Geneva, Switzerland. Experiments at this particle accelerator and others now being upgraded in the United States and Japan are expected to point to new physics beyond the standard model. So the picture I present here for the origin of mass, while being around for almost half a century, may be temporary. In any case, whatever replaces it will probably still be along the same lines.

7.5. ANOTHER WAY TO GET MASS

However, more needs to be said. The masses of the fundamental particles of the standard model constitute only 5 percent of the mass of atomic matter in the universe. The luminous matter in stars is only 0.5 percent. Most of the atomic mass is in nucleons (protons and neutrons). Let's see how that mass comes about.

Nucleons and other strongly interacting particles made of quarks are called *hadrons*. Hadrons made of quark-antiquark pairs are called *mesons*. Examples include pi mesons or *pions* (π) and K-*mesons* or *kaons* (K). The K^+ meson, on which I wrote my doctoral thesis, is composed of an up quark u and an antistrange quark $\bar{s}$. Hadrons made of three quarks are called *baryons*. Examples include the proton ($p = uud$), neutron ($n = udd$), lambda ($\Lambda = uds$), sigma ($\Sigma = uus$), xi ($\Xi^- = dss$), and omega ($\Omega = sss$).

In the standard model, the strong force is described by a theory dubbed *quantum chromodynamics* (QCD). This theory was patterned after the phenomenally successful theory of electromagnetism called *quantum electrodynamics* (QED), developed in the 1940s (postwar), except QCD has three types of charge, called *color charge*, rather than the single electric charge of electromagnetism. And, while electromagnetism in QED is mediated by a single boson, the photon (γ), the strong force in QCD is mediated by eight massless spin 1 gluons (g).

Furthermore, the calculational technique called *perturbation theory* that worked so well for QED is insufficient for precise calculations at low energies. A nonperturbative theory called *lattice QCD* that requires dedicated time on huge supercomputers is used. After three decades since QCD was proposed, computers are now sufficiently powerful to apply lattice QCD to calculate the masses of hadrons. This includes quarks and nucleons.

According to QCD, a hadron is surrounded by a cloud of massless gluons along with some quark-antiquark pairs that flit in and out of the vacuum. These have a total kinetic and potential energy E, which in the rest frame of the hadron will be interpreted as the rest mass $m = E/c^2$.[4] Results from a Japanese supercomputer project called Computational Physics by a Parallel

Array Computer System (CP-PACS) have been remarkable, with the mass spectrum of hadrons calculated with errors of 1 to 2 percent for mesons and 2 to 3 percent for baryons.[5] Quark masses have also been calculated. Since quarks are bound in hadrons, their masses cannot be directly measured and compared with calculations, but the calculated values are consistent with empirical estimates. Even more improved calculations have been under way since this report appeared in 2000.

From this discussion, it should not be surprising that the masses of fundamental particles are relatively small. These masses are basically zero with nonzero values picked up either as a correction needed since the universe is not a pure vacuum in the space between planets, stars, and galaxies, or as the internal energy of a composite system. In any case, these small mass corrections do not call for any fine-tuning or indicate that our universe is in any way special.

8.
Chemistry

8.1. THE HYDROGEN ATOM

Let us take a look at the basic science most directly associated with life, chemistry, which is based on atomic physics. In the nineteenth century, it was observed that the light emitted by gases exposed to a high-voltage electric discharge exhibited a spectrum of wavelengths that was not continuous but marked by sharp lines of very well-defined wavelengths. This could not be explained by the classical wave theory of light. These spectra were found to be different for different substances, and the study of the spectra became an important tool of incredible accuracy for identifying the ingredients in various materials both in the laboratory and far off in space. In fact, the chemical element helium was discovered in the spectrum of the sun before it was found on Earth (hence its name from the Greek *helios*, "sun").

Early in the twentieth century, Niels Bohr was able to calculate the spectrum of hydrogen by using the atomic model proposed earlier by Earnest Rutherford in which a tiny, heavy, positively charged nucleus is orbited in planetary fashion by electrons. Hydrogen is the simplest atom, composed of a single proton as the nucleus and a single electron in orbit around the proton.

Now, electrons in orbit have a problem planets do not. They are electrically charged and should continuously lose energy as they quickly spiral into the nucleus. Bohr postulated that this did not happen because only certain orbits were allowed, those in which the angular momentum was quantized in units of $\hbar = h/2\pi$. Assuming circular orbits of radius r, the electron mass m_e, and electron speed v, the angular momentum L is given by $L = m_e vr = n\hbar$, where $n = 1,2,3,\ldots$.

We call n the *quantum number*.

Sophomore-level physics textbooks all derive from this the energy of each orbit, n,

$$E_n = -\frac{\left(k_E e^2\right)^2 m_e}{2\hbar^2 n^2} = \frac{-13.6\text{eV}}{n^2} \qquad (8.1)$$

The convention is to call zero energy that energy at which the electron is just free of the atom, so its energy is negative when bound to the atom. E_n can be described as a series of *energy levels* as shown in figure 8.1. The lowest energy level at E_1 = –13.6 eV is called the *ground state*. The next highest level at E_2 = $-13.6/2^2 = -3.4$ eV is the *first excited state*.

When a high-voltage electric spark passes through hydrogen gas, the hydrogen atoms get excited to higher energy levels. Then they drop back down to lower levels, emitting photons whose energy equals ΔE, the difference between the energies of the higher and lower levels of the transition.

The wavelength of the emitted radiation is related to the energy of the photons by $\lambda = hc / \Delta E$, given by,

$$\frac{1}{\lambda} = \frac{\Delta E}{hc} = R\left(\frac{1}{n_f^2} - \frac{1}{n_i^2}\right) \tag{8.2}$$

where n_i and n_f are the quantum numbers of the initial and final energy levels, λ is the photon wavelength, and R is the *Rydberg constant*, $R = 1.097 \times 10^7\ \text{m}^{-1}$. This formula was first determined experimentally from spectroscopic measurements in 1888 by the Swedish physicist Johannes Rydberg.

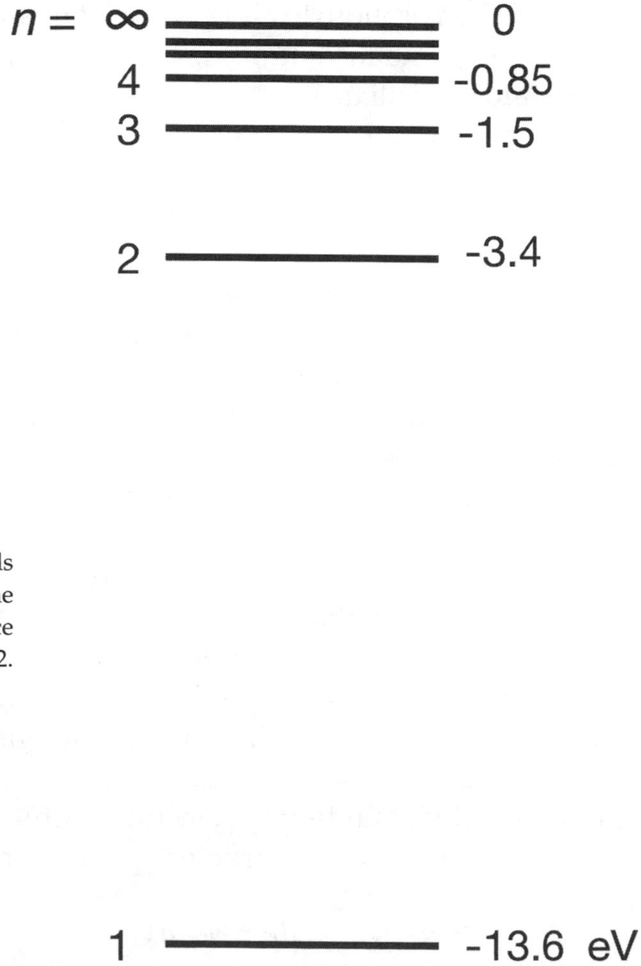

Fig. 8.1. Energy levels of hydrogen. Note the large difference between levels 1 and 2.

8.2. THE QUANTUM THEORY OF ATOMS

The Bohr model of the hydrogen atom is usually referred to as the "old quantum theory." Modern quantum theory was independently created by Werner Heisenberg in 1925 and Erwin Schrödinger in 1926. Their methods were unified, and the theory was placed in a more elegant form by Paul Dirac in 1930. Since Schrödinger's version utilizes the lowest level of mathematical sophistication, partial differential equations, it is the one that most scientists recognize.

Although there is more to all versions of quantum mechanics, the time-independent Schrödinger equation for the hydrogen atom is most familiar:

$$-\frac{\hbar^2}{2m_e}\nabla^2\psi - \frac{e^2}{r}\psi = E\psi \qquad (8.3)$$

where ∇^2 is the *Laplacian operator* of differential calculus and $\psi = \psi(\mathbf{r})$ is the *wave function*, which is a function of the vector $\mathbf{r}$ (magnitude r) from the proton to the electron. This differential equation can be solved to give the same energy levels (8.1) derived by Bohr. In addition, one obtains the wave function, which is in general a complex number.

The picture of the atom in the "new" quantum theory is quite different than that proposed by Rutherford and Bohr. Instead of well-defined, planetary-like orbits around the proton, the electron is visualized as a cloud in which its actual position is uncertain (even, if you will, undefined).

The probability of finding the electron within an infinitesimal volume dV centered at a particular position $\mathbf{r}$ is given by

$$dP = |\psi|^2 \, dV \qquad (8.4)$$

8.3. THE MANY-ELECTRON ATOM

Atomic physicists and chemists use the Schrödinger equation to calculate the structures of the atoms beyond hydrogen that constitute the chemists' *Periodic Table of the Elements* found on the wall of every chemistry classroom. The mathematics cannot be solved exactly beyond helium, so numerical methods must be used. However, although we cannot calculate the structures precisely, those structures are still determined by the Schrödinger equation. It's just like the solar system, where we cannot accurately compute planetary orbits, but those orbits are still determined by the law of gravity.

The Schrödinger equation for a general atom can still be written down and its properties studied. Consider the atom with atomic number Z, which is the number of protons in the nucleus and the number of electrons surrounding the nucleus when the atom is electrically neutral. Non-neutral atoms, called *ions*, can be handled the same way and need not be discussed here.

Let $\mathbf{r}_i$ be the position vector of electron i and $\mathbf{R}_j$ be the position vector of proton j. Then, taking into account the electric force between all charged particles and neglecting the motions of the protons and neutrons in the nucleus, which are very slow compared to the electrons, Schrödinger's equation becomes

$$\left\{-\frac{\hbar^2}{2m_e}\sum_{i=1}^{Z}\nabla_i^2-\sum_{i=1}^{Z-1}\sum_{j=i+1}^{Z}\frac{k_E e^2}{|\mathbf{r}_i-\mathbf{R}_j|}+\sum_{i}^{Z-1}\sum_{k=i+1}^{Z}\frac{k_E e^2}{|\mathbf{r}_i-\mathbf{r}_k|}\right\}\psi=E\psi \quad (8.5)$$

This is complicated, but it is completely explicit. It describes every atom in the periodic table. Note that I have taken care of not only the electrical force between each electron and each proton in the nucleus but also the forces between each electron pair. The neutron has a small magnetic interaction with the electron and protons, but this can be neglected for most chemical purposes.

8.4. THE SCALING OF THE SCHRÖDINGER EQUATION

Let us discuss how the Schrödinger equation depends on the fundamental constants m_e and α.[1]

First, rewrite the Schrödinger equation so that distances are expressed in units of the *Bohr radius,*

$$r_B = \frac{\hbar^2}{k_E m_e e^2} = \frac{\hbar}{m_e c \alpha} \tag{8.6}$$

That is, define new electron coordinates $\mathbf{s}_i = \mathbf{r}_i / r_B$ and proton coordinates $\mathbf{S}_i = \mathbf{R}_i / r_B$. Substituting these in (8.5), we get

$$\left\{ -\frac{1}{2}\sum_{i=1}^{Z} \nabla_i^2 - \sum_{i=1}^{Z}\sum_{j=1}^{Z} \frac{1}{\left|\mathbf{s}_i - \mathbf{S}_j\right|} + \sum_{i=1}^{Z-1}\sum_{k=i+1}^{Z} \frac{1}{\left|\mathbf{s}_i - \mathbf{s}_k\right|} \right\} \psi = \frac{E}{\alpha^2 m_e c^2} \psi \tag{8.7}$$

where

$$\alpha = \frac{k_E e^2}{\hbar c} \tag{8.8}$$

is the *fine structure constant*, the dimensionless strength of the electromagnetic force.

Note that from (8.1) the binding energy of the hydrogen atom can be written more transparently as

$$E_1 \equiv E_B = -\frac{1}{2}\alpha^2 m_e c^2 \tag{8.9}$$

We conclude: *the structure of atoms is independent of any fundamental parameters.*

All m_e and α do is change the scale of energy. All universes with a wide range of values of these parameters will have the same chemistry. This will change only when we can no longer assume the electrons reside outside the nucleus or when the motion of protons can no longer be neglected. Certainly m_e and α are not fine-tuned to give chemistry and life based on it.

8.5. RANGE OF ALLOWED PARAMETERS

Let us estimate the limits on parameters for which chemistry remains unchanged. Consider an atom of atomic number Z that has been stripped of all its electrons except one so that it is hydrogen-like.

The nucleus will have Z protons, so the lowest orbit of the electron will be given by the Bohr radius divided by Z,

$$a_Z = \frac{\hbar}{Zm_e c\alpha} \tag{8.10}$$

As we start adding electrons to the atom, they screen each other somewhat from the nucleus so that the effective atomic number becomes Z' where $Z' < Z$.

The lowest orbit then becomes

$$a_{Z'} = \frac{\hbar}{Z'm_e c\alpha} \tag{8.11}$$

so that $a_{Z'} > a_Z$. Thus we can take a_Z given by (8.10) to be the minimum radius of an electron in an atom of atomic number Z.

Whatever the value of Z', the lowest orbit electron will always be higher than that for the single electron atom because of the screening.

The radius of the nucleus will depend on the number of nucleons in the nucleus A (the mass number), the proton mass

m_p, and the strong interaction strength α_S. The condition for chemistry to be independent of any fundamental parameters, then, is that the lowest electron orbital radius of a single-electron atom with Z protons in the nucleus will be greater than the radius of any nuclei with Z protons, that is, any value of A.

$$R_A = \frac{\hbar}{\alpha_S m_p c} A^{1/3} \tag{8.12}$$

Condition: $a_Z > R_A$ or, from (8.10) and (8.12),

$$\frac{\hbar}{Z m_e c \alpha} > \frac{\hbar}{\alpha_S m_p c} A^{1/3} \tag{8.13}$$

and

$$\beta = \frac{m_e}{m_p} < \frac{1}{A^{1/3} Z} \frac{\alpha_S}{\alpha} \tag{8.14}$$

The result places an upper limit on the ratio $\beta = m_e / m_p$, ranging from 16.4 for $_1H^1$ to 0.0154 for $_{92}U^{237}$, for the values of $\alpha_S = 0.12$ and $\alpha = 1/137$ in our universe. In our universe, $\beta = 5.45 \times 10^{-4}$. So, holding α_S / α constant, β can be 28 times bigger and anything smaller. This same limit will hold for α; it can be 28 times its existing value of 1/137 and anything smaller. Similarly, $1/\alpha_S$ can be 28 times its existing value of 8.3 and anything smaller. Furthermore, allowing all the parameters to vary, we see that there is plenty of parameter space for chemistry to be chemistry.

What is more, we can argue that the electron mass is going to be much smaller than the proton mass in any universe even remotely like ours. Recall our discussion of the origin of mass in chapter 7. The electron gets its mass by interacting electroweakly with the Higgs boson. The proton, a composite particle, gets most of its mass from the kinetic energies of gluons swirling around inside. They interact with one another by way of the strong interaction, leading to relatively high kinetic energies. Unsurprisingly, the proton's mass is much higher than the electron's and is likely to be so over a large region of parameter space.

In short, we will have the same chemistry over wide ranges of the parameters α, α_S, and β. None are fine-tuned for chemistry as we know it.

9.

The Hoyle Resonance

9.1. MANUFACTURING HEAVIER ELEMENTS IN STARS

In the standard cosmological model, the nuclei of lighter elements, H^2 (D), He^3, He^4, and Li^7, are produced by what is called *hot big-bang nucleosynthesis* in the early universe. All the naturally occurring heavier elements are produced in stars by *stellar nucleosynthesis*.

After a star has burned all its hydrogen, it begins to contract, as it no longer has the pressure to withstand gravity. This results in increasing temperature, and then other nuclear processes occur that manufacture the heavier nuclei. If the star is sufficiently massive, greater than ten solar masses, it will end its life in a giant explosion, called a *supernova*, that distributes the heavy nuclei into space, where they pick up electrons and form the elements of the periodic table. The white dwarfs that mark the final stage of less massive stars also contribute by evaporating off heavy elements. Planets then form as gravity gathers these elements into clumps.

Without stellar nucleosynthesis, the universe would not have the ingredients for life. The most important element for our form of life is carbon. The basic mechanism for the manufacture of carbon is the fusion of three helium nuclei into a single carbon nucleus, the so-called *triple-alpha* or *3α* process (the He^4 nucleus is the α particle that constitutes α radiation):

$$3He^4 \rightarrow C^{12}$$

However, the probability of three bodies coming together simultaneously is very low, and some catalytic process in which only two bodies interact at a time must be assisting. An intermediate process had earlier been suggested in which two helium nuclei first fuse into a beryllium nucleus, which then interacts with the third helium nucleus to give the desired carbon nucleus:

$$He^4 + He^4 \rightarrow Be^8$$
$$He^4 + Be^8 \rightarrow C^{12}$$

Other nuclei important to life can also be formed from He^4 fusing with other nuclei:

$$He^4 + C^{12} \rightarrow O^{16}$$
$$He^4 + O^{16} \rightarrow Ca^{20}$$

The He^4 nucleus is the most tightly bound of the nuclei and is treated as a basic unit in these processes. The other nuclei are also tightly bound, with the exception of Be^8, which is unstable, decaying back into two α's.

9.2. THE HOYLE PREDICTION

As we saw in chapter 2, one of the more famous of the claimed anthropic coincidences is the existence of an excited state of the carbon nucleus at an energy where sufficient carbon is produced in stars to make carbon-based life possible. In physics parlance, this is called a "resonance," since it acts like a tuning fork when it produces an amplified effect.

In 1953, astronomer Fred Hoyle calculated that the production of carbon by means of the Be^8 catalyst would not occur with sufficient probability unless that probability was boosted by the presence of an excited nuclear state of C^{12} at a very specific energy. Hoyle proposed that this previously unknown state must exist at about 7.7 MeV.[1] The existence of such a state was quickly confirmed experimentally.[2]

This has been regarded by theists as a miraculous example of fine-tuning. They like to quote Hoyle himself:

> A commonsense interpretation of the facts suggests that a superintellect has monkeyed with physics, as well as with chemistry and biology, and that there are no blind forces worth speaking about in nature. The numbers one calculates from the facts seem to me so overwhelming as to put this conclusion almost beyond question.[3]

For many authors not interested in theological consequences, Hoyle's prediction is still widely regarded as the first (and only) successful application of what is called "anthropic reasoning." This is an inference made about the natural world that is based on the fact of the existence of life. Recently, Danish philosopher Helge Kragh has gone back through the history of the Hoyle resonance and questioned whether the prediction was, in fact, the result of anthropic reasoning. She found that Hoyle did not originally connect the resonance with the existence of life, an association that was not made until much later in the 1980s. She concludes, "from a historical point of view it is misleading to label the prediction of the 7.65 MeV state anthropic or to use it as an example of the predictive power of the anthropic principle."[4]

Physicists as notable as Nobel laureate Steven Weinberg

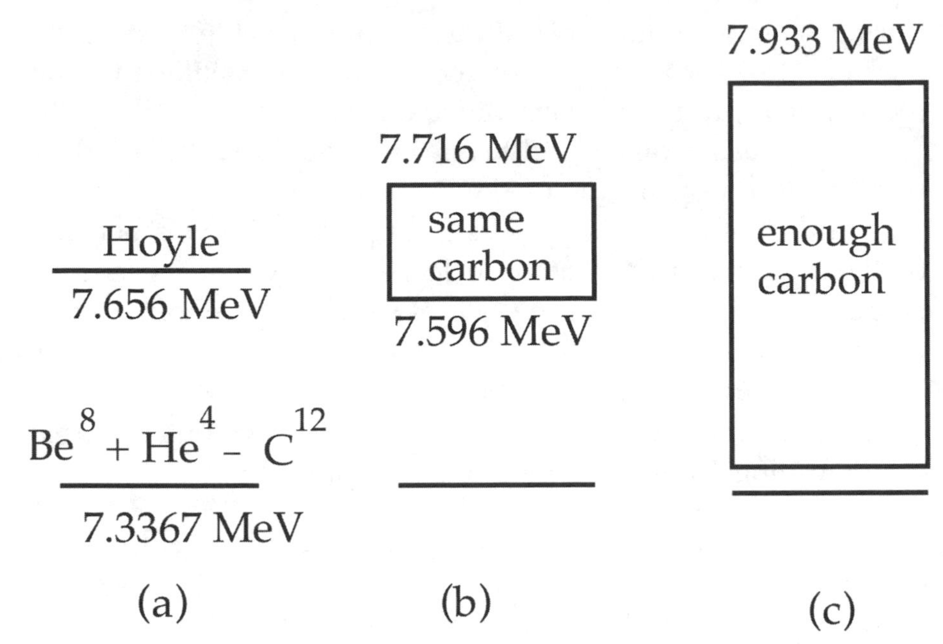

Fig. 9.1 (a) Energy levels of the carbon nucleus showing the ground state and the excited state predicted by Hoyle; (b) the allowed range of the excited state that would yield the same amount of carbon; (c) the range of energy for an excited state that would produce carbon adequate for life.

have questioned how fine-tuned this excited state really is.[5] Let us look at the details.

Figure 9.1(a) shows two energy levels: (1) the amount by which the total rest energy of Be^8 + He^4 exceeds that of the C^{12} nucleus, which is 7.3367 MeV; (2) the excited state of C^{12} predicted by Hoyle and observed at 7.656 MeV. On this scale, the ground state of C^{12} is zero.

Now it is often claimed that this excited state has to be fine-tuned to exactly this value in order for carbon-based life to exist. This is not true. This is just the energy level that gives the C^{12} abundance in our universe. Life might be possible over a range of energies.

In 1989, physicist Mario Livio and his collaborators made calculations to test the sensitivity of stellar nucleosynthesis to the exact position of the excited C^{12} state predicted by Hoyle.[6]

They determined that a 0.06 MeV increase in the location of the level to 7.716 MeV would not significantly alter the carbon production in stellar environments. A decrease by the same amount to 7.596 MeV was needed before the carbon production increased significantly above its value in our universe. This range is shown in figure 9.1(b). Already we can see the excited state is not very fine-tuned.

Finally, the problem is not to get the exact amount of carbon in our universe but simply sufficient carbon production for life. We get more carbon when the Hoyle energy level is even lower. Furthermore, Livio et al. showed that the energy level can be increased by as much as 0.277 MeV to 7.933 MeV before insufficient carbon is produced. As figure 9.1(c) shows, an excited state anywhere from this energy down to near the minimum energy would produce adequate carbon.

Hoyle also pointed to another apparent example of fine-tuning. The O^{16} nucleus has an energy level at 7.1187 MeV. This is *below* the total energy of $He^4 + C^{12}$, 7.1616 MeV, so it does not effect the reaction rate of these two nuclei. If the O^{16} level had been just a bit higher, most of the carbon would be eaten up producing oxygen.

There is no easy formula one can write down to give nuclear energy levels as a function of fundamental parameters such as α, α, and m_e. But we can still provide a plausible explanation for this fortuitous arrangement of nuclear energy levels that does not require the assumption of divine fine-tuning.

With the current values of the fundamental parameters, all the nuclei that we can view as combinations ("chains") of He^4 nuclei, or α particles, are stable—with the exception of $Be^4 = 2\alpha$. This means that the 2α system is loosely bound and free to vibrate, producing harmonic oscillator energy levels or "resonances." Here's how Weinberg explains the nuclear dynamics:

> We know that even–even nuclei have states that are well described as composites of α-particles. One such state is the

> ground state of Be^8, which is unstable against fission into two alpha particles. The same α–α potential that produces that sort of unstable state in Be^8 could naturally be expected to produce an unstable state in C^{12} that is essentially a composite of three α particles, and that therefore appears as a low-energy resonance in α–Be^8 reactions. So the existence of this state doesn't seem to me to provide any evidence of fine-tuning.[7]

Weinberg does not discuss the fortuitous absence of a similar state in O^{16} that would result in carbon being destroyed in producing oxygen. However, physicist Craig Hogan expands on Weinberg's picture to include an explanation for the absence of this state,

> We conjecture that the fact that the resonances appear slightly above the sum of the rest masses of the α constituents in these two nuclei [Be^8 and C^{12}] is also a natural result of a symmetry rather than a tuning. It is a natural consequence if the nuclei in these resonant states roughly resemble nonrelativistic oscillators of α particles of fixed mass with small binding energies. In that case, the oscillator energy naturally lies slightly above the summed rest masses of its constituents. In this view, the presence of a resonance in about the right place to enhance nuclear reactions reflects quantum mechanics of the nuclei rather than a fine-tuned balance of forces or masses. Note that because ground-state C^{12}, unlike Be^8, is significantly bound, one would not in this view expect a similarly placed resonance in O^{17}.[8]

In short, the 2α and 3α systems are loosely bound and can oscillate to produce low-energy resonances. On the other hand, the 4α system, tightly bound because of its symmetry, does not produce similar resonant states.

So a good case can be made that no fine-tuning was necessary to produce sufficient carbon for life by way of stellar nucleosynthesis. In chapter 11, I will show that cosmologies are possible in which all the elements are produced primordially in a cold big bang with no need for the Hoyle resonance or any stellar nucleosynthesis. While these cosmologies are probably incorrect for our universe, with its parameters, they might be possible with a change in parameters. Remember, to defeat the fine-tuning argument, I do not have to give a reason why each parameter has the value it does, I must only show that life could be plausible under a wide range of parameters.

10.

Physics Parameters

10.1. ARE THE RELATIVE MASSES OF PARTICLES FINE-TUNED?

As we saw in chapter 7, the fact that the gravitational force between a proton and an electron is 39 orders of magnitude weaker than the electrical force allows for a long-lived universe. This ratio is so small because the masses of the electron and the protons are so small compared to the Planck mass, which is the only "natural" mass you can form from the simplest combination of fundamental constants. I argued that, in the standard model, the fundamental particles all start off with zero mass and then pick up small corrections by way of their interactions with the field of Higgs bosons that pervades the universe and, in the case of strongly interacting particles (hadrons), with the gluon field that surrounds them.

Now I want to consider the claims that the relative masses of the basic particles that make up the familiar atomic matter of the universe are fine-tuned in such a way as to make life possible. These particles are the proton, the neutron, and the electron.

Let us refer to figure 10.1, which shows a possible range of the neutron-proton mass difference. For the moment, let us hold the electron mass fixed. Starting at the bottom of the figure, if the proton were heavier than the sum of the masses of the neu-

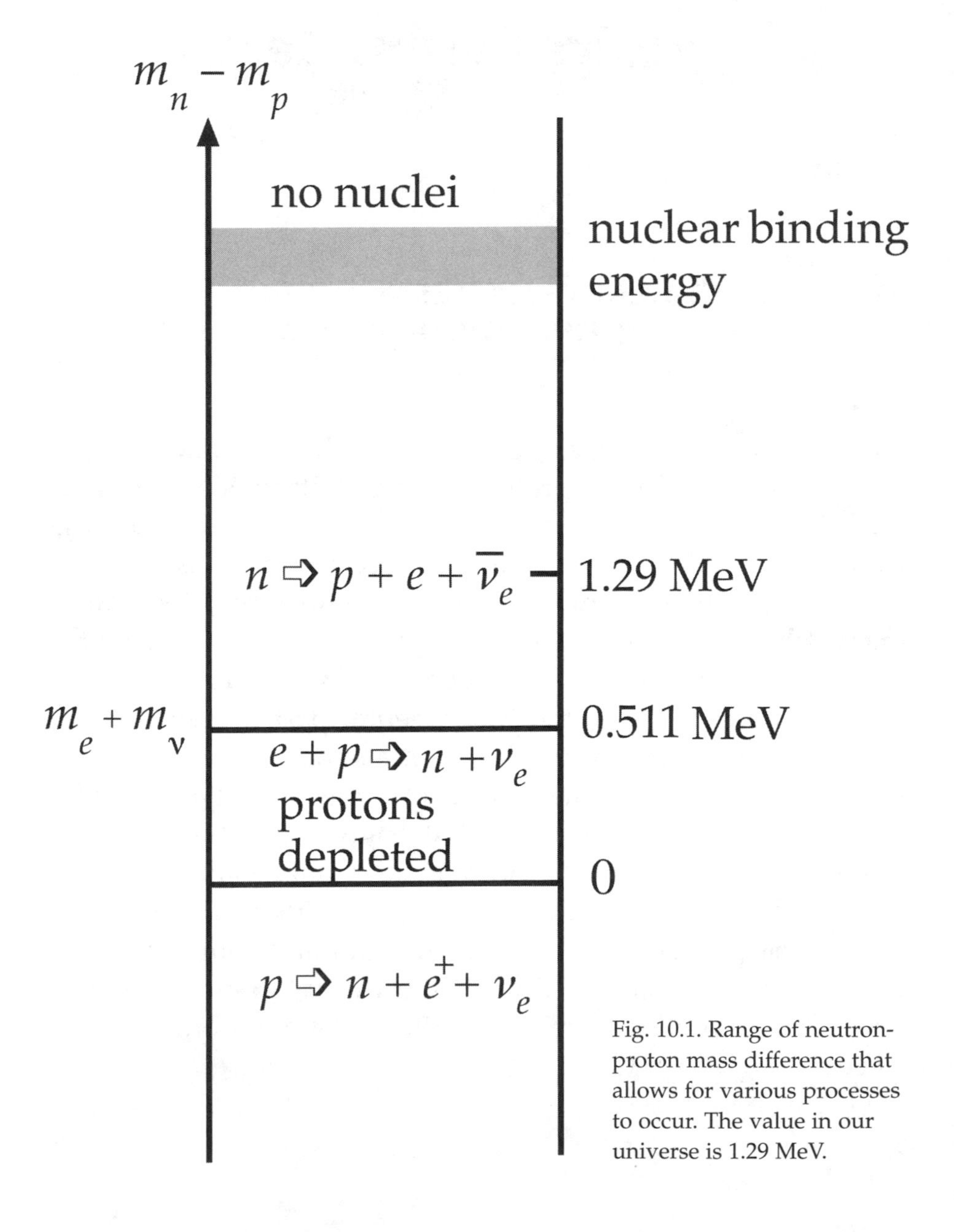

Fig. 10.1. Range of neutron-proton mass difference that allows for various processes to occur. The value in our universe is 1.29 MeV.

tron, positron (anti-electron), and electron neutrino (the neutrino mass is negligible for this purpose), the proton would decay into these three bodies at a relatively high rate (though still a weak interaction) by the process

$$p \rightarrow n + e^{+} + \nu_{e}$$

This would leave few, if any, protons remaining to form hydrogen, an essential element in the universe as we know it.

If, on the other hand, the neutron were heavier than the proton but the mass difference were less than the sum of the masses of an electron and a neutrino, protons would be depleted by the reaction

$$e^{-} + p \rightarrow n + \nu_{e}$$

Moving to the next level, if the *n-p* mass difference is greater than the sum of the masses of an electron and a neutrino, as it is in our universe, with a value of 1.29 MeV, we have the observed neutron decay

$$n \rightarrow p + e^{-} + \bar{\nu}_{e}$$

Finally, we need to understand why, except for a few β–radioactive nuclei, the neutrons inside nuclei do not decay. The requirement here is that the binding energy be greater than the mass difference between the neutron and the sum of the proton and electron masses, that is, 0.781 MeV. Since binding energies result from the strong interaction and are typically on the order of 10 MeV per nucleon, no fine-tuning was needed to make most nuclei stable to β–decay.

Now, the above discussion was just based on mass differences. The actual abundances of the particles in the universe will also depend on the cosmological situation, which affects reaction rates, as will be described in section 10.4 below.

This would seem to be a particularly good example of fine-tuning. But this apparently fortuitous arrangement of masses has a plausible explanation within the framework of the standard model. As we saw in chapter 7, the proton and neutron get most of their masses from the strong interaction, which makes no distinction between protons and neutrons. If that were all there was to it, their masses would be equal. However, the masses and charges of the two are not equal, which implies that the mass difference is electroweak in origin.

The proton is composed of two *u* quarks and one *d* quark, while the neutron is one *u* quark and two *d* quarks. Again, if quark masses were solely a consequence of the strong interaction, these would be equal. Indeed, the lattice QCD calculations discussed in chapter 7 give the *u* and *d* quarks masses of 3.3 ± 0.4 MeV.[1]

On the other hand, the masses of the two quarks are estimated to be in the range 1.5 to 3 MeV for the *u* quark and 2.5 to 5.5 MeV for the *d* quark.[2] This gives a mass difference range $m_d - m_u$ from 1 to 4 Mev. The neutron-proton mass difference is 1.29 MeV, well within that range. We conclude that the mass difference between the neutron and proton results from the mass difference between the *d* and *u* quarks, which, in turn, must result from their electroweak interaction with the Higgs field. No fine-tuning is once again evident.

10.2. MASS OF ELECTRON

Now, we still have the issue of the electron mass. While its value in our universe is 0.511 MeV, it could have been anywhere from 0 to 1.29 MeV in our universe and 0 to around 4 MeV in any universe and still allow for neutron decay, as illustrated in figure 10.1. The electron mass is much smaller than the proton mass because it gets its mass solely from the electroweak Higgs mechanism, so being less than 1.29 MeV is not surprising and also shows no sign of fine-tuning.

Note also that as we allow all the masses to vary, not just one at a time, we will have plenty of parameter space for nuclei to form. If the neutron-proton mass difference were less than the current electron mass of 0.511 MeV, a smaller electron mass would give us nuclei. If the electron mass were greater than the current neutron-proton mass difference of 1.29 MeV, a large *n-p* mass difference would compensate for that.

Let me add a short comment on quark masses. Robert L. Jaffe and two collaborators have published an extensive study that they call "an environmental impact statement" on the masses of quarks. They allow up to three quarks to participate in nuclei, fixing the mass of the electron and the lightest baryon. They classify as *congenial* those worlds that satisfy the environmental constraint that the quark masses allow for stable nuclei with charge one (hydrogen), six (carbon), and eight (oxygen), making organic chemistry possible. For two light quarks with charges 2/3 and –1/3 of the unit electric charge, they find a band of congeniality 29 MeV wide in mass difference, with our own world lying comfortably away from the edges.[3]

10.3. MASSES OF NEUTRINOS

Here's one that did not trip up Hugh Ross but was added to the list by Rich Deem (his fine-tuned parameter number 30). Deem claims that if the neutrino masses (remember there are three types) were smaller, then galaxy clusters, galaxies, and stars would not form. If they were larger, galaxy clusters and galaxies "would be too dense."[4]

Deem assumes that the number of neutrinos in the universe is fixed. It is not. Neutrinos form part of the background left over from the big bang, along with the more familiar photons in the cosmic background radiation. Both form gases of free (noninteracting) particles with fixed total energy.

The cosmic neutrino background is composed of all three

types of neutrinos. If their total energy is E, the total number of neutrinos will depend on their masses. Decrease the masses, and the number increases; increase the masses, and the number decreases. However, the effective total mass E/c^2 stays the same, and so the net gravitational effect is independent of neutrino mass.

10.4. STRENGTH OF WEAK NUCLEAR FORCE

Ross claims that if the strength of the weak nuclear force α_W were larger, too much hydrogen would convert into helium in the big bang. There then would be too much heavy element material made by stars. If α_W were smaller, too little helium would be produced and stars would convert too little matter into heavy elements, making the chemistry of life impossible.

Let us look at the role played by the weak interaction in the early universe. When the temperature is above roughly 1 MeV,[5] an equilibrium between elementary particles is maintained by the reactions

$$n \leftrightarrow p + e^- + \bar{\nu}_e$$

$$\nu_e + n \leftrightarrow p + e^-$$

$$e^+ + n \leftrightarrow p + \bar{\nu}_e$$

which are all weak interactions.[6]

The weak interaction strength parameter α_W affects the rates of the above reactions, which also depend on temperature. In order to maintain equilibrium, the reaction rates must exceed the expansion rate of the universe given by the Hubble parameter H.[7] When the temperature falls below some temperature T_F, the reaction rates fall below H and the neutron-to-proton ratio freezes at a certain ratio n/p that depends on the n-p mass difference and the temperature at freeze-out.

$$\left(\frac{n}{p}\right)_{freeze-out} = \exp\left(-\frac{m_n - m_p}{T_F}\right) \tag{10.1}$$

For our universe, $m_n - m_p = 1.293$ MeV, $T_F = 0.72$ MeV, and $n/p = 1/6$ (using energy units where $c = k_B = 1$).[8] Neutron decay reduces n/p to about $1/9$.

Recall the discussion in section 10.1 where it was suggested that protons would be depleted if the *n-p* mass difference were less than the sum of the masses of an electron and a neutrino since there then would be no neutron decay. This could happen if the electron were more massive. This would eliminate the reaction $n \rightarrow p + e^- + \bar{\nu}_e$, but the others would still occur, with essentially the same result: a net surplus of protons over neutrons of 6 to 1.

The higher the value of α_W, the faster the rates and the earlier they freeze out. In that case, T_F is higher, and the ratio n/p at freeze out is higher. We see that the greater number of protons results from their lower mass, which I have already justified. Most of the neutrons are incorporated into He^4 when the average kinetic energy is about 0.1 MeV, with others going to deuterium, Li^7, and trace amounts of other light elements.

If the value of $(m_n - m_p)/T_F$ is too large, there will be fewer neutrons to combine into light nuclei. If $(m_n - m_p)/T_F$ is very small, then we might have equal numbers of protons and neutrons, and with most combining into helium, we would be left with few free protons to make hydrogen.

However, n/p does not have to be exactly $1/6$ or $1/7$ for life to be possible, just as $m_n - m_p$ does not have to be exactly 1.293 MeV. For purposes of argument, let me assume that a range $1/10 < n/p < 1/3$ is viable. Then values of $m_n - m_p$ and T_F with the band shown in figure 10.2 would be acceptable. Once again we see there is no reason to assume fine-tuning.

How important is the weak interaction anyway? In 2006, Roni Harnik and two collaborators published a paper in which

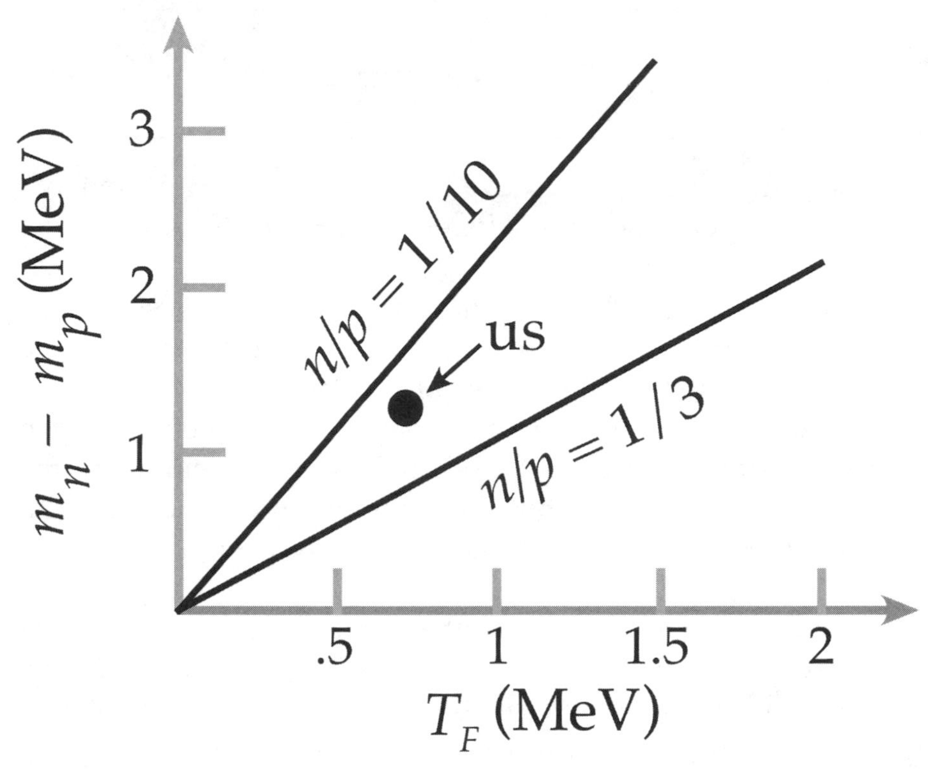

Fig. 10.2. The range of neutron-proton mass difference, $m_n - m_p$, and freeze-out temperature, T_F, for which the neutron-to-proton ratio ranges from 1/3 to 1/10. T_F is determined by the weak interaction strength α_W.

they considered a universe that contained no weak interactions.[9] They refer to four papers where it is shown that a very narrow range of α_W is allowed for a universe to have our kind of life. However, each analysis fixed all parameters except α_W. Harnik et al. found that they could produce livable universes by varying other parameters over a wide range.

Here is their procedure:

> We define a habitable universe as one having big-bang nucleosynthesis, large-scale structure, star formation, stellar burning through fusion for long lifetimes (billions of years) and plausible means to generate and disperse heavy elements

> into the interstellar medium. As a consequence, we will demand ordinary chemistry and basic nuclear physics be largely unchanged from our Universe that matching to our Universe is as straightforward as possible. We are not aware of insurmountable obstacles extending our analysis to planet formation, habitable planets, and the generation of carbon-based life. Nevertheless, these questions are beyond the scope of this paper, and we do not consider them further.[10]

The viability of a weakless universe is disputed in an unpublished paper by physicists Louis Clavelli and Raymond White.[11] They argue that life will be strongly inhibited in a universe without weak interactions since insufficient oxygen would be produced and oxygen is critical to our form of life in many ways. They also mention a number of other problems, such as the importance of parity violation, a product of the weak interaction, in producing left-handed DNA and the role of CP violation in giving us the asymmetry between baryons and antibaryons in the universe.

Clavelli and White have followed the lead of the theistic supporters of fine-tuning in assuming that any form of life must necessarily be very much like our own. Thus we need oxygen for water, which is needed as a universal solvent. I e-mailed Harnik, who commented by return e-mail that the claim that life needed a universal solvent is "not based on evidence." While it is true for our form of life, "that's not a large sample." In any case, he says, "even in the weakless universe there can be plenty of water."[12]

Life may someday be found elsewhere in our universe. I am sure it's out there, given how easy it is to produce complex molecules from simpler stuff. In 1953, graduate student Stanley Miller did it in the laboratory in a few weeks.[13] I will wager that extraterrestrial life, sufficiently distant to not be connected to Earth life, will not be based on left-handed DNA. So why make that a requirement for life in this and every universe? I expect life to occur in any sufficient system in any sufficiently long time, and our kind of biology is not a constraint.

Still, it is hard to conceive of a universe without weak interactions because of the fundamental role they play in the standard model. I think I can make the case against fine-tuning without relying on the Harnik et al. proposal. Indeed, many of my explanations are given within the framework of the standard model, so I need not start changing it.

10.5. STRENGTH OF THE STRONG INTERACTION

One of Martin Rees's "six numbers" that determine the structure of the universe he calls *nuclear efficiency*.[14] It is defined specifically as ε = the fraction of the mass of helium that is greater than the mass of two protons and two neutrons.

$$\varepsilon = \frac{M_{He} - 2\left(m_p + m_n\right)}{M_{He}} \qquad (10.2)$$

The value in our universe is $\varepsilon = 0.0010$. It is a measure of the stickiness of the glue that holds a nucleus together and is ultimately determined by the dimensionless strong interaction strength parameter α_S. If the glue isn't sticky enough to overcome the electrical repulsion of the two protons, the nucleus will be unstable. Rees asserts that if $\varepsilon < 0.006$, deuterium and many other nuclei would not exist.

On the other hand, if $\varepsilon > 0.008$, the proton-proton repulsion is insufficient to keep protons from sticking together. In that case, you would have no free protons and no hydrogen. Instead, two protons would bind together directly to produce He^2, helium without any neutrons in the nucleus, which is unstable in our universe.

This is a beautiful example of how many of the fine-tuning arguments can be misleading. Fine-tuners make the common mistake, and it is purely and simply a mistake, of holding everything fixed and varying just a single parameter. Let's see what happens when the electromagnetic strength α is also a free parameter: If ε is too small to overcome the electrical repulsion of protons, then we can simply lower the value of α, the electromagnetic strength, thereby weakening the repulsion and allowing the nuclei to stick together. Let's assume, for illustration, a linear relationship between ε and α place an upper bound on ε above which there are no free protons and a lower bound where nuclei are unstable, as shown in figure 10.3.

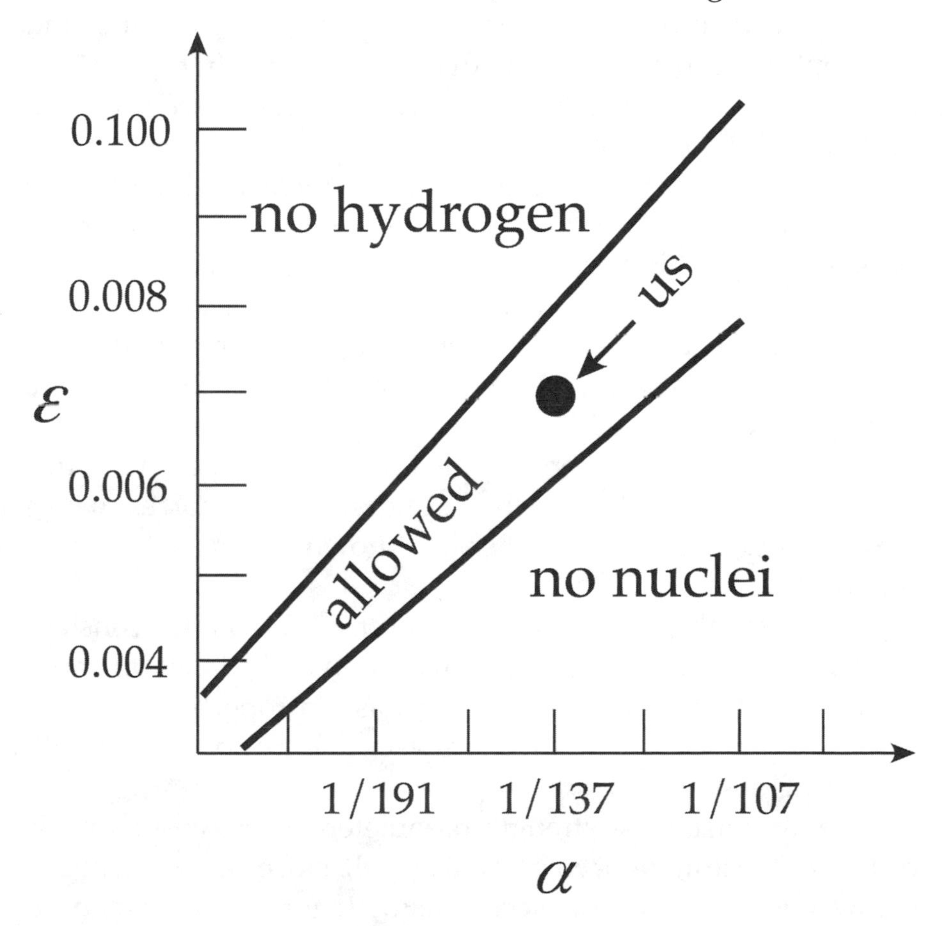

Fig. 10.3. The parameter space for ε and α showing the allowed region for nuclear stability and free protons.

We see that when we allow more than one parameter to vary, fine-tuning is no longer evident. The parameter ε is not fine-tuned to the range 0.006 to 0.008 but can take on a wide range of values if α is also allowed to vary.

10.6. THE RELATIVE STRENGTHS OF THE FORCES

In this and previous chapters I have shown that most of the claimed fine-tuned parameters have natural ranges of values within known physics that allow plenty of space for some kind of complex life to be formed with no fine-tuning. In this section, let us look at the strength of the various forces to see if we can understand their ranges of values.

The strength of the electromagnetic force is given by the dimensionless parameter $\alpha = k_E e^2 / \hbar c$, where k_E is Coulomb's constant. This is historically known as the *fine-structure constant*, since it first appeared when physicists began examining the small splittings of spectral lines that were not accounted for in the simple Bohr model of the atom. It has the well-known value usually written as 1/137.

However, this value only holds for low energies, and α, like the strength parameters of the strong and weak forces, varies with energy. This is a fact either not known or ignored by most fine-tuners.

The strength parameter of the strong force is not constant even at low energies, ranging from around 0.6 at 1 GeV to about 0.1 at 1000 GeV. The strong force has a property known as *asymptotic freedom*, in which it grows weaker and weaker with increasing energy.

The dimensionless strength parameter for the weak force is difficult to estimate accurately by itself since, as we will see below, it is related to the electromagnetic force in the standard model, which contains electroweak unification. It is about $\alpha_W \approx 10^{-7}$ at 1 GeV.

In chapter 7 we saw that the dimensionless gravitational strength parameter α_G was *arbitrarily* defined as $Gm^2 / \hbar c$, where m_p is the proton mass 1.67×10^{-27} kg. With this definition, the relative strength of gravity to electromagnetism is

$$\frac{\alpha_G}{\alpha} = \frac{\dfrac{Gm_p^2}{\hbar c}}{k_E \dfrac{e^2}{\hbar c}} = \frac{Gm_p^2}{k_E e^2} \quad (10.3)$$

8.09×10^{-37}, which many authors have deemed so very unlikely that it has to be fine-tuned. However, as we discussed in chapter 7, the value of α_G is arbitrary. The reason the gravitational force seems so weak compared to the electromagnetic force is that the masses of elementary particles are small compared to the Planck mass. No fine-tuning is required. Small masses are a natural consequence of the origin of mass. The masses of elementary particles are essentially small corrections to their intrinsically zero masses. The correction is larger for hadrons, such as the quark and the proton, than it is for leptons, such as the electron and the neutrino, which do not interact strongly. This results from the fact that the strength parameter of the strong interaction, α_S, is greater than that of the weak and electromagnetic interactions, α_W and α. And why is this so?

The answer lies partially with the standard model of particles and forces and probably will not be fully answered until we reach the next level of understanding. However, for my purposes I do not need to know that theory. I just have to show that we have a plausible explanation for the force strengths, and the burden is on the theist to show why these are necessarily fine-tuned.

The standard model is a theory that is fundamentally gauge symmetric (point-of-view invariant) with the electromagnetic and weak forces united as the electroweak force. The strong force is also gauge symmetric and operates alone. The dimensionless strengths of the forces are given by three dimensionless numbers, α_1, α_2, and α_3. Because of symmetry breaking in the electroweak force, α_1 and α_2 are each mixtures of α and α_W.

$$\alpha_i = \frac{g_i^2}{4\pi} \quad i = 1,2,3 \tag{10.4}$$

where the "electroweak coupling constants" are $g_1 = e / \cos\theta_W$ and $g_2 = e / \sin\theta_W$, and θ_W is the Weinberg or weak mixing angle that measures the amount of broken symmetry. Natural units $\hbar = c = 1$ are conventionally used.

Note that

$$\frac{1}{\alpha} = \frac{1}{\alpha_1} + \frac{1}{\alpha_2} \tag{10.5}$$

The quantities α, θ_W, and α_S are all determined by experiment. The value of θ_W is about 28 degrees.[15]

The strong force is not broken at pre–Large Hadron Collider (LHC) energies and $\alpha_S = \alpha_3$. Theories in which the electroweak and strong forces are unified are called *grand unified theories* (GUTs), but none of these are established empirically.

In the standard model, the force strengths have a specified energy dependence that is linear on a log plot for the inverse strengths. If you plot α_i^{-1} to higher energies in the conventional model, they converge, but not on the same point, as seen by the dashed lines in figure 10.4. However, in an extended version of the standard model, MSSM—the *minimum supersymmetric standard model*—where *supersymmetry* treats bosons and fermions equally, the three parameters converge on a single point, as seen by the solid lines.[16]

Three random lines converging on a single point over twelve orders of magnitude is very unlikely, so this seems to be

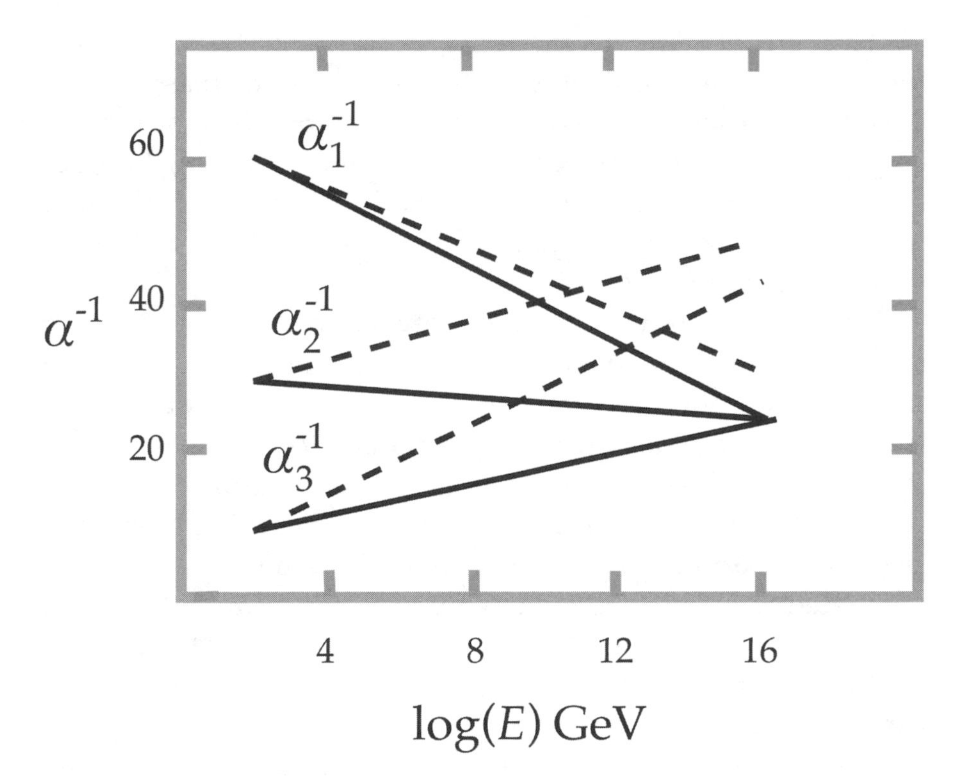

Fig. 10.4. The inverses of the force-strength parameters in the standard model (dashed lines) and in the standard model extended to include supersymmetry (solid lines) as a function of energy. The latter converge at a common energy 3×10^{16} GeV. At this *unification energy*, all three forces are equal.

an impressive prediction. It is a promising sign that supersymmetry will be an important part of the new physics beyond the standard model that physicists will be exploring shortly, as the LHC has now gone into operation.

What is the significance of this result for the fine-tuning question? All the claims of the fine-tuning of the forces of nature have referred to the values of the force strengths in our current universe. They are assumed to be constants, but, according to established theory (even without supersymmetry), they vary with energy. They are also always assumed to be independent, and we see they are not.

The current standard model agrees with all the data to date (pre-LHC), which is below 1 TeV in center-of-mass energy. According to scientific cosmology, the temperature of the universe was 1 TeV at 10^{-12} second after the start of the big bang. So we can safely say we understand (that is, have models that successfully describe) the physics of the universe back to when it was only a trillionth of a second old.

This allows us to view the universe as starting out in a highly symmetric state with a single, unified force. Prior to 10^{-37} second, when the temperature of the universe was above 3×10^{16} GeV, the single force has a strength $\alpha_U = 1/25$. At 10^{-37} second, when the temperature of the universe dropped below 3×10^{16} GeV, symmetry breaking separated the unified force into electroweak and strong components with strengths α_1 and α_2 associated with the electroweak force and $\alpha_3 = \alpha_S$ representing the strong force. These force strengths diverged from one another as time progressed, making for a more complex universe and thus one more likely to have some form of life built on complexity. The electroweak force became weaker than the unified force, while the strong force became stronger. At 10^{-12} second, when the temperature of the universe dropped below 1 TeV, the breaking of gauge symmetry caused the two electroweak components to mix into the familiar electromagnetic and weak forces. In short, the parameters will differ from one another at low energies, but not by orders of magnitude.

So we see that the relation between the force strengths is natural and predicted by the highly successful standard model, supplemented by the yet unproved but highly promising extension that includes supersymmetry. If this turns out to be correct, and we should know in few years, then it will have been demonstrated that the strengths of the strong, electromagnetic, and weak interactions are fixed by a single parameter, α_U, plus whatever parameters are remaining in the new model that will take the place of the standard model.

10.7. PROTON DECAY

Hugh Ross includes proton decay in his list of fine-tuned phenomena. He argues that if the decay rate of the proton were greater, life would be exterminated by the release of radiation. If it were smaller, there would be insufficient matter in the universe for life.[17]

Protons, neutrons, and other composite hadrons with half-integer spin, such as the hyperons ($\Lambda, \Sigma, \Xi, \Omega$), are called *baryons* and are assigned *baryon number* $B = 1$. They are composed of three quarks, where each quark has $B = 1/3$. *Mesons* are hadrons composed of quark-antiquark pairs and have zero baryon number, where the antiquarks have $B = -1/3$. Leptons, such as the electron and the neutrino, contain no quarks and so also have zero baryon number. Antibaryons, such as the antiproton and the antineutron, are composed of three antiquarks and so have a net baryon number $B = -1$. To this day, observations are consistent with the principle of *baryon number conservation*. Baryon number is conserved in reactions such as

$$p + \bar{p} \rightarrow \pi^+ + \pi^-$$

Leptons, the electron, the muon, the tauon, and three neutrinos, are assigned lepton number $L = 1$. Each has an associated antilepton with $L = -1$. Observations are also so far consistent with the conservation of lepton number, as in

$$e^+ + e^- \rightarrow \gamma + \gamma$$

Photons and the other gauge bosons have $B = 0$ and $L = 0$.

If the universe came from nothing or tunneled from a previous universe, it would be expected to have net baryon and lepton numbers zero and an exact balance between baryons and antibaryons, and between leptons and antileptons. They might have quickly annihilated one another, as in the above reactions, leaving behind just photons and neutrinos and nothing to make the galaxies, had not something happened. Early in the universe, baryon number conservation or lepton number conservation, or both, must have been broken, leaving a slight excess of protons to capture electrons and to otherwise interact to make hydrogen and the light elements that eventually formed stars.

The violation of baryon number conservation is predicted by grand unified theories that unite the electroweak and strong interactions. It would result in the decay of the proton, the most likely reaction being

$$p \to e^{+} + \pi^{o}$$

In figure 10.5, we see the Feynman diagram for this particular proton decay process.

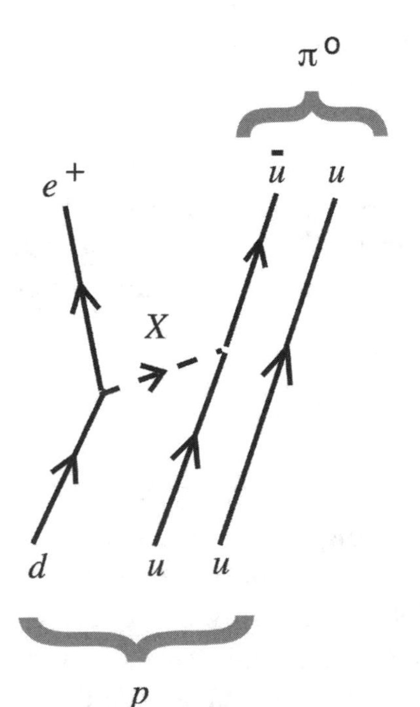

Fig. 10.5. Proton decay into a positron and a neutral pion, as predicted by grand unified theories (GUTs), but not yet observed in experiments. The X is a gauge boson of the advanced theory, not part of the standard model, which allows for the violation of baryon number and lepton number conservation. Note that the X has charge –4/3 for the direction shown.

The decay rate is roughly given by

$$\Gamma_p = \alpha_U^2 \frac{m_p^5}{M_X^4} \tag{10.6}$$

where α_U is the dimensionless strength of the grand-unified force and M_X is the mass of the gauge boson that mediates the interaction.[18]

Obviously the decay rate has to be very low, or we would not exist. In fact, the simplest GUT, minimal SU(5), was falsified in the 1980s by the failure of its prediction that protons would decay with a mean lifetime of 10^{31} years. The current experimental limit on the mean lifetime (reciprocal of decay rate) for the most likely proton decay, $p \rightarrow e^+ + \pi^0$, is 1.6×10^{33} years.[19] Despite the failure to observe proton decay (yet), any future theory of unified forces is expected to contain this process or something analogous.

We have seen that, while free neutrons decay via β–decay, this is prevented from happening in most nuclei by energy conservation. So we should take into account neutron decay by processes similar to proton decay. Neutrons inside a nucleus decaying by $n \rightarrow e^+ + \pi^-$ would not have an energy barrier.

Let us estimate how low the mean lifetime of nucleon decay (including neutron decay) would have to be to "exterminate" life.[20] The *gray* (Gy) is the unit that is used to measure the deposited energy of radiation inside a body. It corresponds to 1 Joule per kilogram. A radiation flux of 10 gray per hour should be sufficiently lethal to kill off all life. This equals 5.5×10^{17} MeV per kilogram per year. Since the nucleon rest energy is 939 MeV, we have 5.8×10^{14} decays per kilogram per year. One kilogram of matter contains 6.0×10^{26} nucleons and so the mean decay rate is 10^{-12} per year. That is, if the mean nucleon decay lifetime were 10^{12} years or less, we would not expect to have life in the universe.[21] The nucleon decay lifetime in our universe is greater than 10^{33}, twenty-one orders of magnitude higher. No fine-tuning there.

Our calculation for the decay rate may not turn out to be correct, but it's the best we have for making at least a crude estimate and to give us an idea of the possible range of values. The decay rate depends on the strength of the unified force α_U and the mass of the gauge boson that mediates the force M_X. The corrected theory is likely to have the same dependence.

In figure 10.6, I have plotted the mean proton decay lifetime, the reciprocal of decay rate, as a function of M_X for three values

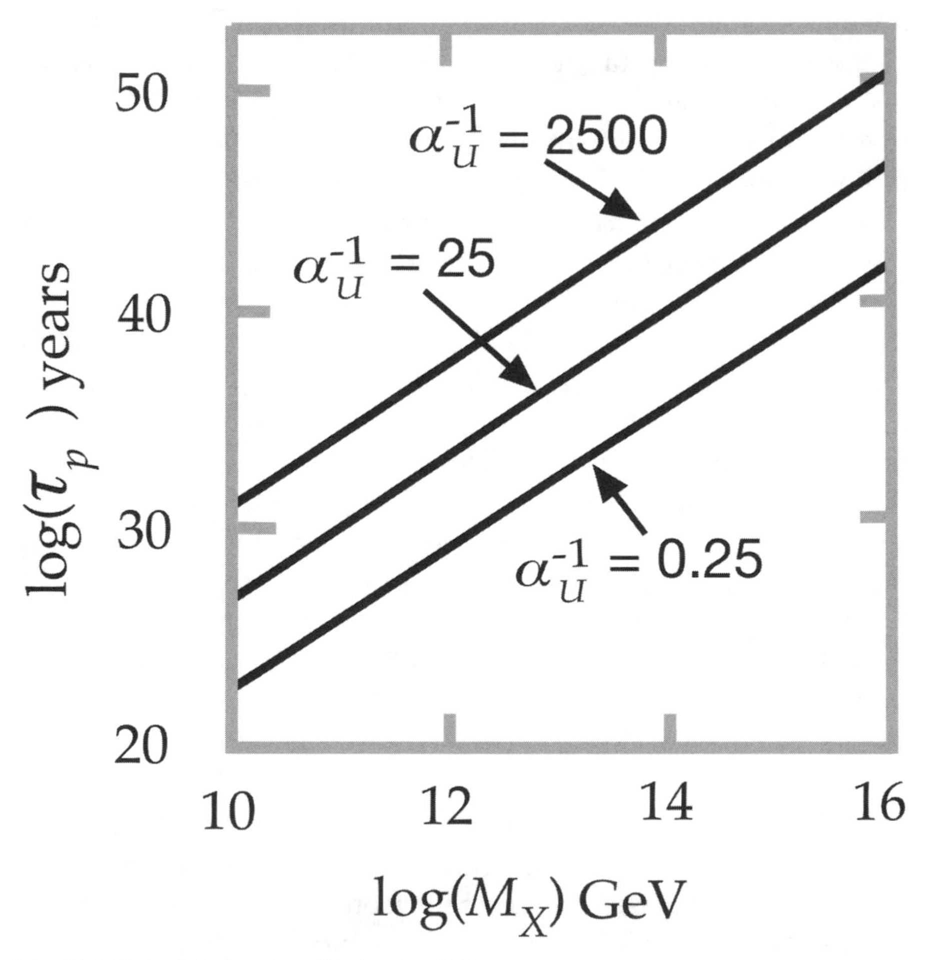

Fig. 10.6. Calculated proton lifetime in GUTs as a function of M_X for three widely spaced values of α_U^{-1}.

of α_U^{-1}. The central line is for $\alpha_U^{-1} = 25$, the value we expect from the extrapolation to the unification energy in figure 10.4. The upper line is for $\alpha_U^{-1} = 2500$, two orders of magnitude higher. The lower line is for $\alpha_U^{-1} = 0.25$, two orders of magnitude lower.

We would have to go another ten orders of magnitude down in lifetime from the lowest point on this graph to reach lethality. Without a doubt, proton decay is not fine-tuned to prevent life being destroyed by radiation.

10.8. BARYOGENESIS

Ross also considered the related question of the excess of matter over antimatter in our universe. He says that if the "initial" excess of nucleons over antinucleons were greater, there would be too much radiation for planets to form; and if there were less, there would be insufficient matter for galaxies or stars to form.[22] These are essentially the same reasons he gave for proton decay (refer to the beginning of the previous section).

It is reasonable to assume that the universe started out with net baryon number $B = 0$, that is, with equal amounts of matter and antimatter. Some processes violating baryon number conservation must have been involved in the early universe. They are present in GUTs, such as the reaction shown in figure 10.5 for proton decay. The generation of a baryon excess in the early universe is called *baryogenesis*.

Back in 1967, the famous Russian physicist and political dissident Andrei Sakharov proposed that three ingredients were necessary for baryogenesis: (1) baryon number violation, (2) C and CP violation, and (3) out-of-equilibrium conditions.[23]

C is the quantum mechanical operator that changes a particle to its antiparticle. CP performs this operation and additionally changes a particle's *handedness* (parity) from left to right or from right to left. When a reaction is indistinguishable from another reaction in which each particle has been replaced by its antiparticle, it is said to be C-invariant. When a reaction is indistinguishable from another reaction in which each particle has been replaced by its antiparticle with its handedness changed, it is said to be CP-invariant. The violation of C invariance was

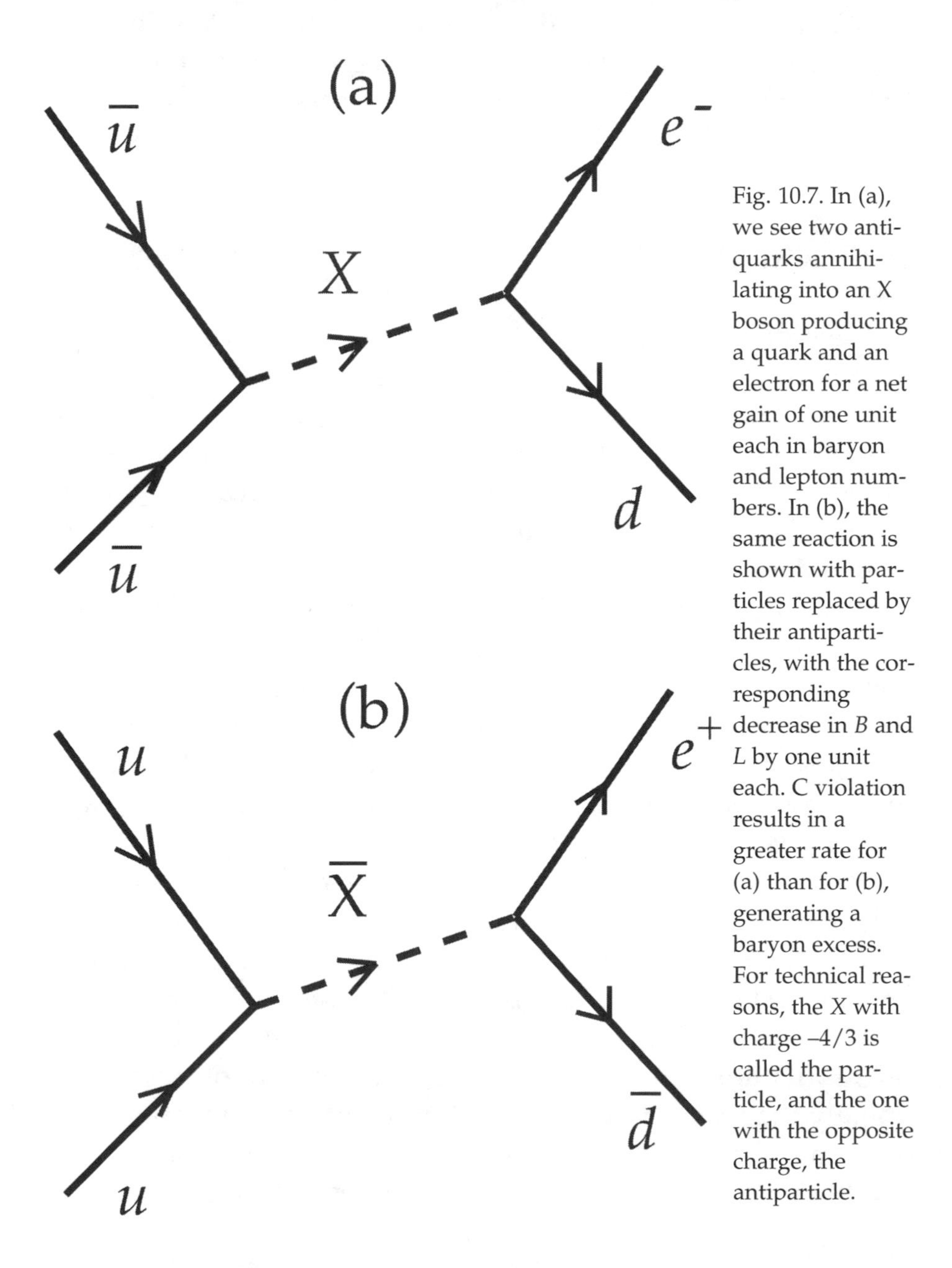

Fig. 10.7. In (a), we see two antiquarks annihilating into an X boson producing a quark and an electron for a net gain of one unit each in baryon and lepton numbers. In (b), the same reaction is shown with particles replaced by their antiparticles, with the corresponding decrease in *B* and *L* by one unit each. C violation results in a greater rate for (a) than for (b), generating a baryon excess. For technical reasons, the *X* with charge –4/3 is called the particle, and the one with the opposite charge, the antiparticle.

seen in 1957, and CP violation was observed in 1964 in neutral kaon decays.

Grand unified theories allow for B, L, C, and CP violation. However, the calculated baryon excess in minimal SU(5) is too small. Since minimal SU(5) is already falsified by the failure of its proton-decay prediction, this is not a contradiction. Unfortunately, this is the only known GUT capable of making quantitative predictions, and we have to await the higher-energy data that will help point the way to a more successful model.

In figure 10.7(a), I show how a reaction involving the X-boson can give a net reduction in baryon number by 1, with the loss of two $B = -1/3$ antiquarks and the addition of one $B = +1/3$ quark. It also adds a net $L = 1$, with the addition of an electron. Note that $B - L$ is conserved. In figure 10.7(b), the particles are replaced by their antiparticles, and we have net change in baryon and lepton number of –1. If C were not violated, the two reactions would cancel. But C violation implies the two have different rates, so we end up with a greater number of particles than antiparticles. Note that this designation is arbitrary; we just call particles the ones in excess and antiparticles the ones in deficiency in our universe.

So, although we do not yet have a theory to calculate exactly the baryon excess, we can see how it can have come about in grand unified theories. It is hard to see how a greater excess of nucleons over antinucleons would result in too much radiation. Suppose there were no antimatter, the greatest excess of nucleons possible. Then there would be no particle-antiparticle annihilations and no radiation.

Furthermore, we have already seen that we can change the parameters for proton decay by orders of magnitude before lethal radiation would be produced. Since the baryon asymmetry comes from basically the same processes, it is unlikely that that conclusion is going to be changed in this case.

11.
Cosmic Parameters

11.1. MASS DENSITY OF THE UNIVERSE

According to Hugh Ross, if the mass density of the universe were slightly larger, then overabundance of the production of deuterium (heavy hydrogen) from the big bang would cause stars to burn too rapidly for life to form. If smaller, insufficient helium from the big bang would result in a shortage of the heavy elements needed for life.[1]

Observations of great precision in the last decade or two have established that the mass density of our universe is very close to the critical density given by equation (5.10) in chapter 5 for the universe to be balanced exactly on the edge of expanding forever or eventually turning around and contracting back down. Now, as fine-tuned as this may seem, there exists a natural explanation. The data now strongly support the inflationary model in which the universe began with an exponential phase of huge expansion taking about 10^{-35} second, during which it grew by fifty to a hundred orders of magnitude. During that time, whatever the initial geometry of the universe may have been, it would have flattened out like the surface of a huge, expanding balloon. Recall from chapter 5 that a universe with flat geometry will have exactly the critical density. This was one of the predictions of the inflationary model that, had it not turned out that way observationally, would have falsified the model. In any case, it is clear that the density of the universe is not fine-tuned. It equals the critical density to the highest degree of precision possible and is likely to be so in any other universe, for solid, fundamental reasons.

11.2. DEUTERIUM

However, although he has misstated the problem, Ross has a point about deuterium abundance. Cosmologists and physicists have known for years now that the natural existence of deuterium, the isotope of hydrogen with two neutrons, is hard to explain. Any that exists in stars would get burned up because it is an efficient ingredient in nuclear fusion as well as very loosely bound and so easily broken apart in high heat. One of the great triumphs of the hot big-bang model was to explain how deuterium is produced by nucleosynthesis in the early universe. No one has been able to think of any other mechanism for deuterium production, and the successful quantitative prediction of deuterium abundance together with those of the other light elements (such as helium and lithium) has helped prove beyond a reasonable doubt that the big bang happened.[2]

Now, the deuterium abundance is very sensitive to the density of matter in the universe. Actually, as we saw in chapter 5, the matter (or energy) of the universe is composed of three terms:

- atomic or "baryonic" matter (4 percent), density ρ_B,
- dark matter (23 percent), density ρ_{DM}, and
- dark energy (73 percent), density ρ_{DE},

where $\rho_B + \rho_{DM} + \rho_{DE} = \rho_c$. This can be expressed in dimensionless form,

$$\Omega_B + \Omega_{DM} + \Omega_{DE} = 1 \tag{11.1}$$

where each term is the density parameter, the fraction of the critical density of each component, $\Omega = \rho / \rho_c$.

The deuterium abundance depends only on the density of baryonic matter, which is mainly composed of neutrons and protons. Figure 11.1 shows the calculated abundance of the number of deuterium nuclei (one proton and one neutron) relative to ordinary hydrogen (a single proton) by big-bang nucleosynthesis plotted as a function of Ω_B. The horizontal gray band gives the measured value of the abundance.[3] The vertical gray band gives the measured value for Ω_B. The widths of the bands

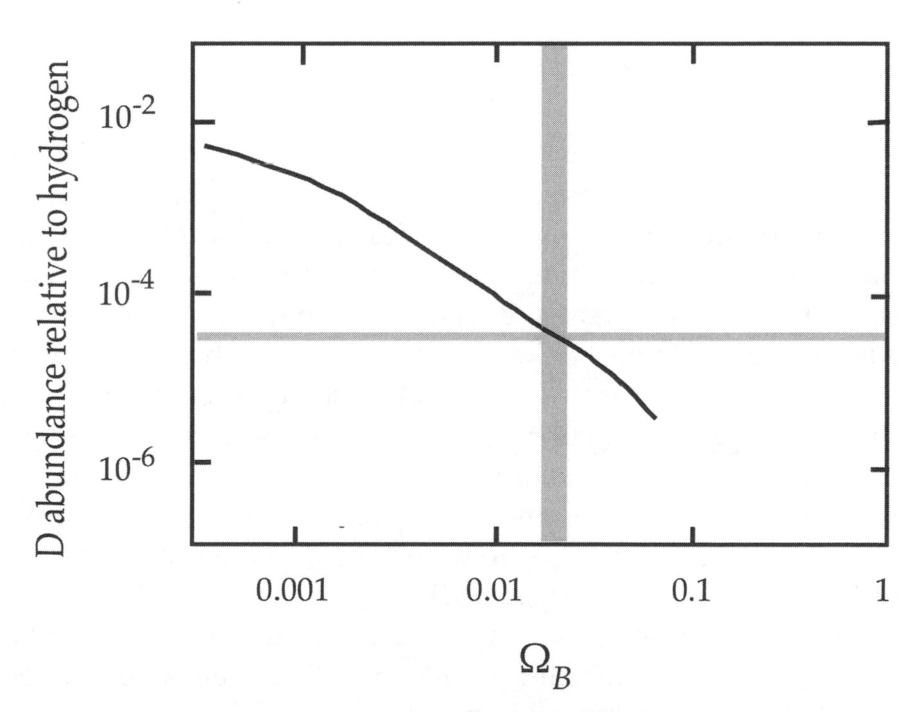

Fig. 11.1. Deuterium abundance by mass as a function of the density parameter for baryons. The gray bands give the measured values and errors.

indicate the small measurement error on each. The theory prediction is right on.

Now, it is certainly true that deuterium abundance depends sharply on Ω_B, but it is hardly fine-tuned. Our universe has a deuterium abundance of about 2×10^{-5}, which isn't much. Clearly not a lot of deuterium is needed for life. Ω_B can vary by two orders of magnitude, from 0.001 to 0.1, and still produce relative amounts of deuterium from 10^{-6} to 10^{-2}. Ross has claimed that too much deuterium would cause stars to burn too rapidly.[4] I have not seen any calculation that demonstrates this. The deuterium nucleus is so loosely bound that large amounts are unlikely to survive the high temperatures and the turbulence of the early big bang to accumulate in stars. Ω_B can vary by two orders of magnitude and still produce the small amounts of deuterium needed for a living universe.

11.3. THE EXPANSION RATE OF THE UNIVERSE

As I mentioned in the preface, in a number of his debates, William Lane Craig has said, "Stephen Hawking has estimated that if the rate of the universe's expansion one second after the Big Bang had been smaller by even one part in a hundred thousand million million, the universe would have re-collapsed into a hot fireball."[5] In his book *Life after Death*, Dinesh D'Souza used the same allusion: "If the rate of expansion one second after the Big Bang had been smaller by even one part in a hundred thousand million million, the universe would have collapsed before it ever reached its present size."[6] Neither Craig nor D'Souza give the reference, but I have found the quotation on page 121 of Hawking's *A Brief History of Time*.[7]

The two apologists have again lifted the quote out of context to support their own ill-founded beliefs. A few pages later, Hawking explains how this problem is solved in inflationary cosmology. On page 121 of *A Brief History of Time*, Hawking talks about the inflationary model of the early universe, which was still relatively new in 1988, when his book was published. In the inflationary picture, Hawking notes:

> The rate of expansion of the universe would automatically become very close to the critical rate determined by the energy density of the universe. This could then explain why the rate of expansion is still so close to the critical rate, without having to assume that the initial rate of expansion of the universe was very carefully chosen.[8]

Let me show how that comes about.

The fractional rate of expansion of the universe is called the *Hubble parameter*. In 1929, astronomer Edwin Hubble discovered that the universe was not a firmament, as asserted in the Bible, but was expanding. His measurements indicated that the galaxies were receding from one another with an average speed v that was proportional to the distance between them r, $v = Hr$, where H was the proportionality factor more commonly known as the *Hubble constant*. This was called *Hubble's law*. The universe was getting bigger as time went on.

Think of the chunks of matter given off in an explosion. You would expect that, after some time, those moving faster would be more distant from the point of the explosion than the slower ones, as given by Hubble's law. Suppose the expansion has been going on for a time t. Then, $v = r/t$ and $t = r/v = H^{-1}$.

While the big bang is not quite analogous to an explosion, since it has no special point in space we can call its center but space itself is expanding, we can still use H^{-1} to estimate the age of the universe.[9] That is, the age of the universe is given by the reciprocal of the Hubble parameter, which you get by measuring the slope of the plot of v against r for the galaxies in the universe. The best current value is H^{-1} = 13.7 billion years.

Actually, we now know that the age of the universe is not given precisely by H^{-1}. In 1998, two independent research collaborations discovered that the recessional speed of galaxies is increasing with distance. The expansion of the universe is accelerating. In fact, as described in chapter 5, the universe appears to be going through an inflationary expansion just as it did 13.7 billion years ago, only far more slowly.

We will get to this later. The effect is small, and, for our present purposes, we can take $H^{-1} = a / v$ as approximately equal to the age of the universe.

In chapter 5 and again earlier in this chapter, we saw that the average density of the matter in the universe has now been precisely measured to be close to the critical density given by (5.10). This means that the Hubble parameter, the fractional rate of expansion of the universe, is

$$H = \sqrt{\frac{8\pi G\rho_c}{3}} \qquad (11.2)$$

Thus, ρ_c is determined by H, which could still take on a range of values.

However, it wouldn't matter much whether the universe is 13.7 billion years old, or 12.7 or 14.7, so it is hardly fine-tuned. If the universe were only 1.37 billion years old, then life on Earth or elsewhere would not yet have formed; but it might eventually. If the universe were 137 billion years old, life may have long ago died away; but it still could have happened. Once again, the apologists' blinkered perspective causes them to look at our current universe and assume that this is the only universe that could have life, and that carbon-based life is the only possible form of life.

In any case, it is clear that the expansion rate of the universe is not fine-tuned to "one part in a hundred thousand million million."

11.4. PROTONS AND ELECTRONS IN THE UNIVERSE

Hugh Ross claims that if the ratio of the number of protons to the number of electrons in the universe, n_p / n_e, were larger, electromagnetism would dominate over gravity and galaxies would not form. If smaller, gravity would dominate and chemical bonding would not occur.[10]

The argument here assumes the ratio is some arbitrary constant that could have had any value except for God's intervention. In fact, the number of electrons exactly equals the number of protons for a very simple reason: as far as we can tell, the universe is electrically neutral, so the two particles must balance because they have opposite charge. No fine-tuning happened here. The ratio is determined by conservation of charge, a fundamental law of physics.

Here is a clear explanation for how this all came about in the early universe, as explained in *Extragalactic Astronomy and Cosmology*, by astronomer Peter Schneider:

> Before pair annihilation [the time when most electrons annihilated with antielectrons, or *positrons*, producing photons] there were about as many electrons and positrons as there were photons. After annihilation *nearly* all electrons were converted into photons—but not entirely because there was a very small excess of electrons over positrons to compensate for the positive electric charge density of the protons. Therefore the number density of electrons that survive the pair annihilation is exactly the same as the number density of protons, for the Universe to remain electrically neutral.[11]

Note that if the universe came from nothing, its total charge should be zero, as expected if there were no miraculous creation.

11.5. BIG BANG RIPPLES

Martin Rees identifies a dimensionless parameter *Q* that represents the amount of energy needed to break apart and disperse stars, galaxies, and clusters of galaxies that are held together by gravity, in proportion to the rest energy mc^2 of the object.[12] In our universe, *Q* is on the order of 10^{-5}, indicating gravity is relatively weak and the objects are easy to break apart.

Q is a measure of the lumpiness of matter. A completely smooth universe would have $Q = 0$. While the lumpiness is only one part in 100,000, we still need to explain where this lumpiness came from. Until the inflationary model came along, cosmologists had no convincing idea on how to account for the formation of galaxies, clusters, and stars.

In the inflation model, the universe starts out very small, so quantum effects are important. The energy density of the vacuum, which is associated with the cosmological constant, will not be smooth but will fluctuate in accordance with the uncertainty principle. While inflation considerably smoothed out these primordial fluctuations, a small amount of variation in the temperature of the cosmic background radiation remained when inflation came to a halt and the Hubble expansion commenced. An observed variation of one part in 100,000 in the temperature variation of the cosmic background radiation was reported by the Cosmic Background Explorer experiment (COBE) in 1992 and has since been confirmed by other, even more precise, observations. This is precisely the *Q* value mentioned above.

While heroic attempts by the best minds in cosmology have not yet succeeded in calculating the magnitude of *Q*, inflation theory successfully predicted the angular correlation across the sky that has been observed.[13]

Indeed, the great uniformity of the cosmic background across the sky is difficult to explain without inflation providing the causal connections needed. Suppose we look at the radiation from two opposite points on the celestial sphere, each 10 billion light-years from Earth. Since they are 20 billion light-years apart, and the universe is only 13.7 billion years old, it would seem that at the time when the photons left their sources they were still separated from one another by 11.3 billion light-years. Yet, when we measure the radiation spectra from both sources, we find they are identical.

Inflation explains this by postulating that the two sources were at the same place and same part of the same system when the original photons were emitted. During inflation, the distance between the sources increased to 11.3 billion light-years, after which they continued to separate by the normal Hubble expansion.

Rees argues that *Q* is fine-tuned to within an order of magnitude of its measured value of 10^{-5}.[14] Many other competent physicists and cosmologists agree, including Paul Davies.[15] According to Rees and Davies, if *Q* were smaller than 10^{-6}, gas would never condense into gravitationally bound structures. On the other hand, if *Q* were "substantially" larger than 10^{-5}, the universe would be a turbulent and violent place. But is an order of magnitude fine-tuning?

Furthermore, Rees, as he admits, is assuming all other parameters are unchanged. In the first case where *Q* is too small to cause gravitational clumping, increasing the strength of gravity would increase the clumping. Now, as we have seen, the dimensionless strength of gravity α_G is arbitrarily defined. However, gravity is stronger when the masses involved are greater. So the parameter that would vary along with *Q* would be the nucleon mass. As for larger *Q*, it seems unlikely that inflation would ever result in large fluctuations, given the extensive smoothing that goes on during exponential expansion.

11.6. PARAMETERS OF THE CONCORDANCE MODEL

The concordance model has not been around for as long as the standard model of elementary particles and forces. The twenty-six parameters for the particle model listed in chapter 4 have values that are the result of three decades of precision experiments at particle accelerators around the world. The only additions to the parameters in that time have been the masses of the three neutrinos, which were added after neutrinos were shown to have mass in 1998. The standard model did not forbid neutrinos from having mass, and they were set zero to agree with the data prior to 1998.

By contrast, the concordance model is still being perfected. The version published by Max Tegmark, Matias Zaldarriaga, and Andrew Hamilton in 2000 has the following parameters:[16] the optical depth at reionization, five parameters describing the density fluctuations that lead to galaxy formation, and five parameters that give the various contributions of the components of the universe to the density of matter. These components are space-time curvature, vacuum energy, cold dark matter, hot dark matter, and baryons (atomic matter). Cold dark matter is the component of the dark matter composed of massive, nonrelativistic particles. Hot dark matter is the component of the dark matter composed of low-mass, extreme relativistic particles. The fit to the data gives negligible hot dark matter. However, see the following section for a discussion of a model with hot dark matter.

Note that the total density of matter and the expansion rate, two parameters that apologists claim are fine-tuned to incredible precision (see 11.1 and 11.2 above), are not listed as parameters of the model to be fit to the data. They are already assumed in the model to have the critical values given by inflation.

The other parameter of the concordance model that is claimed to be fine-tuned is the vacuum energy density. This will be discussed in chapter 12. Five parameters contribute to the density fluctuations that are responsible for galaxy formation, and the fact that they can be fit to the data—which in this case is mainly the structure of the cosmic background radiation—is a testimony to the incredible precision of the WMAP satellite experiment.[17] Fine-tuners do not know what to make of that and have simplified their claims to the single parameter Q discussed in section 11.5 above and again in 11.7 below.

11.7. THE COLD BIG BANG

The cosmology of the concordance model has two eras: inflation followed by a *hot big bang* (HBB). This is now widely accepted as the standard cosmological model because of its good agreement with observational data.[18] That does not mean that all alternatives have been ruled out of consideration. One alternative that goes back to the earliest days of big-bang models is the cold big bang (CBB).

The original HBB attempted to explain the manufacture of all the chemical elements in terms of big-bang nucleosynthesis in a very hot medium. This attempt failed, which lead to the proposal that the elements above lithium are produced by stellar nucleosynthesis. It also required that a good part of the mass of the universe, now called dark matter and dark energy, be nonbaryonic, as seen in figure 11.1.

The cold big-bang model starts with a universe of particles in thermal equilibrium at zero temperature. An important feature is that η_γ, the ratio of photons to baryons, is small, possibly zero, compared to 10^9 in the HBB. This also means there are negligible amounts of antimatter, since most of the early universe photons in the HBB result from matter-antimatter annihilations.

The near or total absence of photons means that the CBB can produce heavier nuclei in the early universe than the HBB, where they are broken apart by photon collisions. Anthony

Aguirre and others have performed CBB nucleosynthesis calculations with η_γ and η_L, the latter being the ratio of leptons to baryons, as free parameters. They are able to produce an interstellar medium with the same general enrichment of chemical elements (not identical ratios) as HBB plus stellar nucleosynthesis. In particular, carbon, nitrogen, and oxygen are produced in abundance.[19]

Structure formation is also easier than in HBB, with lower η_γ. It can happen even without primordial fluctuations! That is, with $Q = 0$. In general, $Q < 10^{-7}$ will allow structure formation. We saw above that even solid scientists such as Rees and Davies claim that Q is one or two orders of magnitude around 10^{-5}. Furthermore, that is only true with η_γ fixed at 10^{11}. Allow this parameter to vary, and Q can change by *many* orders of magnitude.

Now, if the CBB model is so good, why isn't it the standard cosmological model? There is one good reason: the HBB model gives a natural explanation for the cosmic background radiation. When HBB is put together with inflation, we have the standard model of cosmology. It predicts the current temperature of the cosmic background radiation along with its fluctuation spectrum and provides the best fit to a wide range of cosmological data with the fewest parameters.

Some ad hoc mechanism must be cooked up for CBB to yield this highly thermalized background, isotropic and smooth to one part in 10^5. Aguirre does his best, producing a model where the radiation results from so-called *Population III* objects, which include the earliest stars produced in the universe.[20] These were massive, were short-lived, and ended their lives as spectacular supernovae. This so-called bright phase in cosmic history is observed and provides further evidence that the universe is not static but evolves with time. While the HBB model had been proposed before and judged inadequate, Aguirre claims that such thermalization is possible assuming the existence of conducting iron or carbon "whiskers" in intergalactic dust that act as small antennas that absorb and reradiate electromagnetic waves.

I think it is likely that the standard cosmological model of inflation plus the hot big bang, cosmic concordance, is correct for our universe. However, as Aguirre points out, when you start varying parameters by one or two orders of magnitude, the possibility of different physics arises that still could lead to some form of life. The cold big-bang model shows that we don't necessarily need the Hoyle resonance, or even significant stellar nucleosynthesis, for life. With some different fundamental parameters we may have had a cold big bang and no cosmic background radiation. There is no reason why life requires a universe filled with very low energy photons and neutrinos that do not interact with any of the matter of the universe, including each other.

12. The Cosmological Constant

12.1. VACUUM ENERGY

In chapter 5 it was shown that the cosmological constant Λ is equivalent to a constant energy density.

$$\rho_{\Lambda} = \frac{\Lambda}{8\pi G} \tag{12.1}$$

Since no matter or radiation is associated with the cosmological constant, this corresponds to the vacuum itself having energy. Now, the cosmological constant comes from general relativity, where it is associated with the curvature of space-time. General relativity is not a quantum theory. Yet most physicists somewhat illogically attribute the source of vacuum energy to "quantum fluctuations," in which particles and their antiparticles flicker in an out of existence. That is, they point out that because of the Heisenberg uncertainty principle, energy conservation can be violated for short periods of time, allowing for particles to briefly appear.

The problem is when you actually calculate the energy of a

pair of particles in a vacuum, you find it is zero. The process of particle pair creation and annihilation from zero energy to zero energy in fact conserves energy. Now, it is true that a process called *vacuum polarization* is known to be very important in quantum electrodynamics. There the momentary appearance of an electron-positron pair (which is composed of the electron and positron with equal and opposite charges) forms a tiny electric dipole in the space near an atom and produces a shift in the atom's spectrum (*Lamb shift*) and other minuscule effects. The great triumph of quantum electrodynamics (QED) in the late 1940s was the independent calculation to great precision of these "radiative corrections" by Julian Schwinger, Richard Feynman, and Sin Itiro Tomanaga.[1]

However, this process takes place only in the vicinity of atoms and is not to be confused with the energy of the empty vacuum. A better model is provided by the zero-point energy of particle fields. When you quantize the familiar simple harmonic oscillator, you find that the lowest energy level is not zero but has a value $hf/2$, where f is the frequency of the oscillator. This is usually attributed to the fact that the oscillating mass cannot be brought to rest because of the uncertainty principle. At the bottom of its swing, the mass has a well-defined position and so its momentum cannot be zero.

In quantum field theory, elementary particles are described as the "quanta" or excitations of certain abstract quantum fields. For example, the photon is the quantum of the electromagnetic field. The electron is the quantum of the Dirac field. In general, the quantum field is a multidimensional mathematical function that has a set of values at each point in space-time and accounts for all the particles of a given type, such as photons or electrons, at those points in space-time.

A free (noninteracting) field of fixed energy integer spin particles (bosons) has equally spaced harmonic oscillator energy levels separated by hf, where h is Planck's constant and f is the frequency of the oscillation (see figure 12.1). Each level corre-

boson field | fermion field

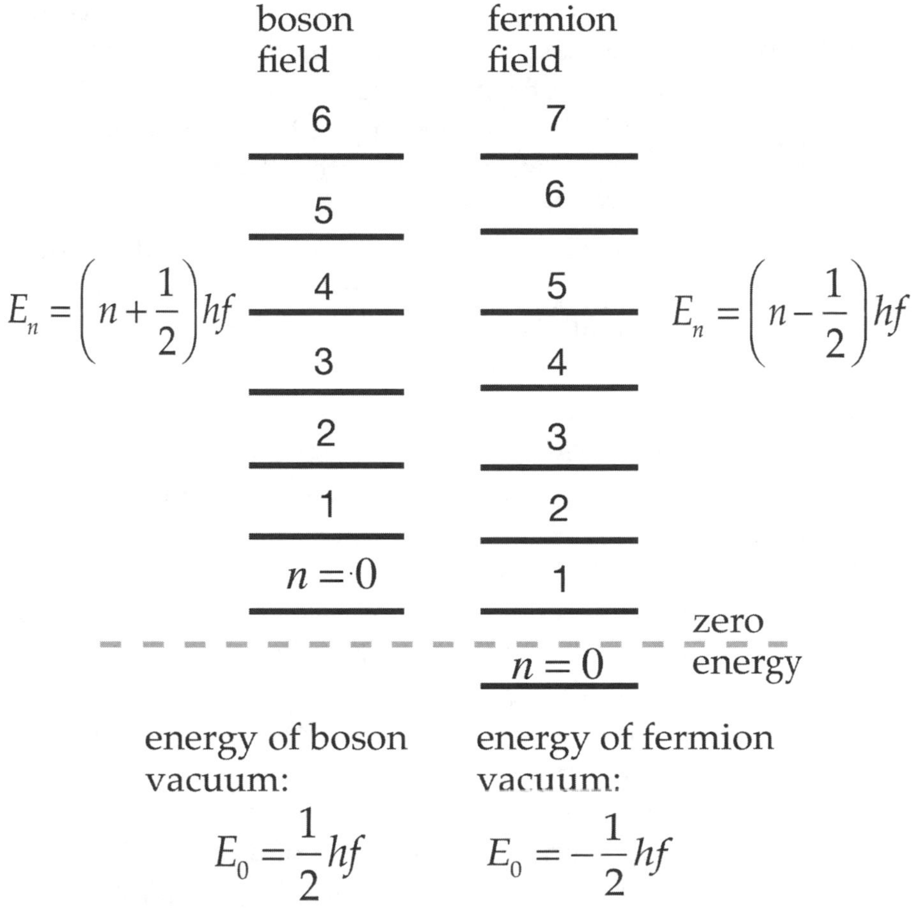

$$E_n = \left(n + \frac{1}{2}\right)hf \qquad E_n = \left(n - \frac{1}{2}\right)hf$$

boson field levels: 6, 5, 4, 3, 2, 1, $n = 0$

fermion field levels: 7, 6, 5, 4, 3, 2, 1, $n = 0$

zero energy

energy of boson vacuum:

$$E_0 = \frac{1}{2}hf$$

energy of fermion vacuum:

$$E_0 = -\frac{1}{2}hf$$

Fig. 12.1. The energy levels of noninteracting fields of fixed energy bosons and fermions. The number of particles, n, in each level is indicated. Each level has one more particle than does the one below. The lowest level for bosons has zero particles but an energy $hf/2$, called the zero-point energy. Fermions have the same levels, except their zero-point energy is $-hf/2$.

sponds to a given number of particles of that energy. The full spectrum of energies is built up as the sum of those levels. The lowest energy level contains zero particles but has a nonzero energy $hf/2$ called the *zero-point energy* that is familiar from the quantized harmonic oscillator. The energy levels of fermion (half integer spin) fields are the same except that the zero-point energy is $-hf/2$.

Now, if it were not for gravity, the zero-point energy could be ignored. In Newtonian physics, you can always add an arbitrary constant to the potential energy of a system without changing the forces involved, since potential energy depends only on energy differences. A similar procedure is possible in quantum field theory when dealing with electromagnetism or the subnuclear forces. However, in general relativity, the geometry at a point in space-time is determined by the energy-momentum tensor that depends on the absolute energy and momentum densities of the distribution of matter at that point. Thus the nonzero energy of the vacuum can have a gravitational effect. This has led people to associate the zero-point energy with the energy of the vacuum and then that with the cosmological constant.

General relativity provides us with no estimate of the value of the cosmological constant. However, we can attempt to calculate the total zero-point energy of the vacuum and hypothetically associate that with the cosmological constant. That calculation involves summing up all the zero-point energy densities of all possible radiation fields. What follows is the standard derivation of the resulting energy density of the vacuum.

In chapter 5, I explained that the state of a classical system of particles is represented by a point in abstract *phase space*. Recall that each axis in phase space represents a degree of freedom of the system. For a single point particle, the phase space axes are its position coordinates (x, y, z) and its canonical momentum components (p_x, p_y, p_z). For N such particles, phase space has $6N$ dimensions. If you don't have simple point particles but bodies with other degrees of freedom, such as rotation and oscillation, we have additional coordinate axes for these.

In quantum mechanics, the uncertainty principle prevents you from measuring the position in phase space with unlimited accuracy and so you must divide the space into cells, as seen in figure 12.2, where the area of each two-dimensional side of the cell is Planck's constant h. For N particles, the cell will have $3N$

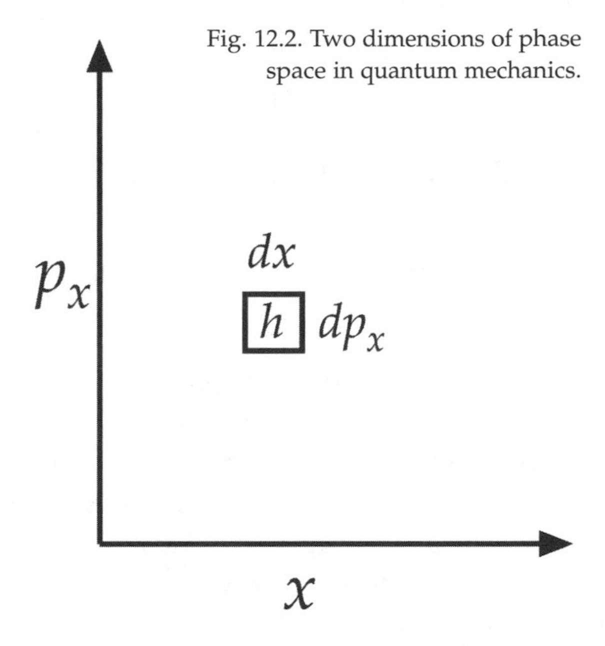

Fig. 12.2. Two dimensions of phase space in quantum mechanics.

sides (not counting opposite sides), and the total phase space volume of the cell will be h^{3N}. In this case, you will have a one-particle state in each volume element in phase space. For an extended body, you will have as many states as there are internal degrees of freedom, g.

The number of states in a cell is

$$dN = \frac{g}{h^3} d^3p d^3x \tag{12.2}$$

In the following, the equations are a lot less cluttered and, I think, the results clearer, if we work in units where $\hbar = c = 1$. Then, $h = 2\pi$ and the three-vector momentum $\mathbf{p} = \hbar\mathbf{k} = \mathbf{k}$, where $\mathbf{k}$ is the wave number. The direction of the three-vector $\mathbf{k}$ is random, so you can integrate over it. The density of states is then defined as the number of states per unit configuration space volume:

$$dn = \frac{4\pi g}{(2\pi)^3} k^2 dk \tag{12.3}$$

The *total* zero-point energy density then is

$$\rho = \pm \frac{1}{2} \int_0^{k_{\max}} \omega dn = \pm \frac{4\pi g}{(2\pi)^3} \frac{1}{2} \int_0^{k_{\max}} \sqrt{k^2 + m^2}\, k^2 dk \tag{12.4}$$

where + is for bosons, – is for fermions.

Since $k = 2\pi/\lambda$, where λ is the de Broglie-Compton wavelength, we should have a maximum k when λ is on the order of the Planck length, smaller distances being operationally indefinable. That is, $k_{max} = M_{Pl} = 10^{19}$ GeV.

Let's consider the special case of photons, where $m = 0$, $\omega = k$, and $g = 2$. Then the integral is easy, giving,

$$\rho_\gamma = \frac{4\pi(2)}{(2\pi)^3} \frac{1}{2} \int_0^{M_{Pl}} \omega k^2 dk = \frac{1}{8\pi^2} M_{Pl}^4 \tag{12.5}$$

By adding the energy in each cell up to a cutoff energy equal to the Planck energy, about 10^{19} GeV, we get an energy density of about 10^{115} GeV/cm^3. If we don't use a cutoff, we get infinite density, and quantum arguments justify the cutoff. This is 120 orders of magnitude larger than the observed upper limit to the vacuum energy density of the universe, $\rho_{vac} = 10^{-5}\,\text{GeV/cm}^3$. Here we have the famous "worst calculation in physics history," being 120 orders of magnitude larger than the observed limit.

Another, simpler, way to look at this result is to note that the universe can be divided into cells of volume L_{Pl}, the smallest definable volume. Each cell will have an uncertainty in momentum or energy of $M_{Pl} = 1/L_{Pl}$ $(\hbar = c = 1)$ given by the uncertainty principle. Thus the energy density will be approximately $M_{Pl}/L_{Pl}^3 = M_{Pl}^4$.

However, the number 10^{120} should not be bandied about so commonly, since it is based on assuming the boson (specifically photon) field alone. Other boson fields will add to this, and fermion fields will subtract from this, since their zero-point energy is negative. Let B be defined as the number of boson fields, including the degrees of freedom of each field, and let F be the corresponding number of fermion fields.

Assume the masses of the particles make a negligible contribution to the integral in (12.4), so we can write

$$\rho_{vac} = (B - F)\frac{1}{8\pi^2} M_{Pl}^4 \tag{12.6}$$

In chapter 10, we saw that there are some positive indications that at the energy that is now beginning to be probed at the Large Hadron Collider (LHC)—about 1 TeV (one trillion electron-volts)—a new symmetry will kick in, called *supersymmetry*. This implies that the cancellation $B - F$ will be exact above that energy, and we can use an energy cutoff of $M_{Pl} = 1$ TeV = 10^3 GeV, rather than the Planck energy, 10^{19} GeV. Then the vacuum energy density is $\rho_{vac} = 10^{51}(B - F) \quad \text{GeV/cm}^3$.

$$\rho_{vac} = (B - F)\frac{1}{8\pi^2} M_S^4 \tag{12.7}$$

This is still over 50 orders of magnitude higher than the observed limit!

Nevertheless, this is no reason to rush out and claim fine-tuning to 50 orders of magnitude, by God or by nature. Any calculation that disagrees with the data by 50 or 120 orders of magnitude is simply wrong and should not be taken seriously. We just have to await the correct calculation. Let me mention several possibilities in the recent literature, including one exciting new idea called *holographic cosmology* that has attracted a lot of attention and is not limited to the cosmological constant problem.

12.2. HOLOGRAPHIC COSMOLOGY

In 1993, physicist Gerard 't Hooft showed that the maximum information stored in a region of space is the surface area of that region in Planck units.[2] This is called the *holographic principle*.[3] It can be seen to result from the fact that the entropy of a black hole depends on its area, and the maximum entropy of a sphere is that of a black hole of the same radius. One speculative consequence is that the universe itself is a hologram. That is, everything within the three-dimensional volume of the visible universe is specified by bits of information stored on its surface.

In the calculation of the vacuum energy density made above, we summed all the harmonic oscillator states within a given volume. The result is approximately equal to the Planck density.

$$\rho_{Pl} = \frac{M_{Pl}^3}{\frac{4\pi}{3} L_{Pl}} = \frac{3}{4\pi} M_{Pl}^4 \qquad (12.8)$$

The holographic principle says we should have just summed the surface area of that volume. Rather than going back and doing the sum over, we can simply note that the total vacuum energy obtained by that sum will equal the order of magnitude of the mass of a black hole the same size as the universe.

With $c = 1$,

$$\frac{4\pi}{3}\rho_{vac}R^3 = M_{BH} = \frac{R}{2G} \tag{12.9}$$

which gives for the vacuum energy density,

$$\rho_{vac} = \frac{3}{8\pi GR^2} = \rho_c \frac{1}{R^2H^2} = \rho_c \tag{12.10}$$

where we recall the critical density,

$$\rho_c = \frac{3H^2}{8\pi G} \tag{12.11}$$

The holographic principle, if true, automatically implies that the vacuum energy density will just equal the critical density, which is (to a good approximation) the density we observe.

Note that although the volume of the universe is varying with time as the universe expands, we have no problem with energy conservation since the expansion is still adiabatic.

From the first law of thermodynamics:

$$\begin{aligned} dQ &= dU + pdV \\ &= \rho dV + pdV \end{aligned} \tag{12.12}$$

which is zero, since $p = -\rho$ for the cosmological constant.

Thus the holographic principle not only solves the cosmological constant problem but also explains why the vacuum energy density approximately equals the critical density. This solution to these problems has not reached a consensus, and it seems almost too good, too simple, to be true.

However, theorists have discovered that this solution to the cosmological problem gives the wrong equation of state for the dark energy, that of ordinary nonrelativistic matter that has the normal attractive gravity; gravitational repulsion is needed to produce the accelerated expansion of the universe. So the nature of the dark energy is still in question. However, at least we can say with some confidence that the calculation of the vacuum energy that gives the 50 to 120 order of magnitude discrepancy with observations is wrong because it incorrectly sums over the volume of a sphere rather than the area.[4]

12.3. GHOST PARTICLES

In 1967, Andrei Sakharov conjectured that a "ghost universe" of negative energy particles exists that has an effect on the curvature of the universe opposite to that of normal particles.[5] In 1984, Andrei Linde proposed that a separate universe of negative energy particles coexists with ours that cancels out the vacuum energy.[6] Since then, several papers have appeared pointing out the equations of relativistic quantum physics that allow for negative energy solutions, in which cases the zero-point energies cancel and give a net zero vacuum energy.[7]

In relativistic physics, the energy and momentum of a particle of mass m are related by $E^2 = p^2 + m^2$, where $c = 1$, which implies $E = \pm\sqrt{p^2 + m^2}$, that is, both positive and negative energies are allowed. The wave functions of relativistic particles have different forms depending on their spins. If we forget about spins, however, then they all have the same dependence on space and time.

For a free (noninteracting) particle, this is

$$\psi(\mathbf{r},t) = A\exp\left[+i(Et - \mathbf{p}\bullet\mathbf{r})\right] + B\exp\left[-i(Et - \mathbf{p}\bullet\mathbf{r})\right] \quad (12.13)$$

This wave function describes a particle of positive energy E moving in the **p**-direction. That is, in the direction of its three-momentum vector. However, there is another solution that is usually discarded because it describes a positive energy particle moving opposite to the direction of its three-momentum vector.

$$\psi(\mathbf{r},t) = A\exp\left[+i(Et + \mathbf{p} \bullet \mathbf{r})\right] + B\exp\left[-i(Et + \mathbf{p} \bullet \mathbf{r})\right] \quad (12.14)$$

In an unpublished paper distributed over the Internet in 2007, physicist Robert D. Klauber showed in detail how including these "supplemental" solutions, which, after all, are present in the mathematics, results in a vanishing of the zero-point energy.[8]

Let me present a nice way of viewing this situation, brought to my attention by my good friend and colleague, Bob Zannelli. Let me write the phase of the wave function for normal particles as $\pm(Et - \mathbf{p} \bullet \mathbf{r})$. As we have seen, the energy E can be positive or negative. In 1928, Paul Dirac associated the negative energy solutions with antiparticles. The antielectron, or *positron*, was discovered in cosmic rays by Carl Anderson in 1932. In 1949, Richard Feynman showed that negative energy electrons going backward in time can be viewed as positive energy positrons going forward in time.[9] That is, the phase of a negative energy electron, $-E$, going backward in time, $-t$, is the same as above. So normal positive energy particles going forward in time are either particles or antiparticles. Note that this can also be viewed as negative energy particles or antiparticles going backward in time, but that is rarely done, although there is no observational difference between the two.

Now, ghost particles have a phase $\pm(Et + \mathbf{p} \bullet \mathbf{r})$. This can be viewed as negative energy particles or antiparticles going forward in time, or as positive energy particles and antiparticles going backward in time.

This is not just a game. We know that physics is symmetric under the transformation $t \to -t$. Since E is conjugate to t, then physics must be symmetric to the transformation $E \rightarrow -E$. In any case, in an E-symmetric universe, the total energy must be zero. If the universe is empty of matter and radiation, then the vacuum energy must be zero.

12.4. IS THE COSMOLOGICAL CONSTANT ZERO BY DEFINITION?

As mentioned above, the energy of the vacuum is generally assumed to be the net zero-point energy of all quantum fields in the universe. But this calculation, like any in physics that do not include general relativity, has an arbitrary energy zero. Only in general relativity, where gravity depends on mass/energy, does an absolute value of mass/energy have any consequence. So general relativity (or a quantum theory of gravity) is the only place where we can set an absolute zero of mass/energy. It makes sense to define zero energy as the situation in which the source of gravity, the energy momentum tensor, and the cosmological constant are each zero.

Einstein's equation in general relativity is usually written

$$G_{\mu\nu} + \Lambda g_{\mu\nu} = T_{\mu\nu} \qquad (12.15)$$

where $G_{\mu\nu}$ is the *Einstein tensor*, Λ is the *cosmological constant*, $g_{\mu\nu}$ is the *metric tensor*, $T_{\mu\nu}$ is the *energy-momentum tensor*, and we are working in units $8\pi G = c = 1$.

Now, in physics we usually write

$$\textit{Field} = \textit{Source}$$

In this case, then, since both $T_{\mu\nu}$ and Λ are sources of the field $G_{\mu\nu}$, we should write

$$G_{\mu\nu} = T_{\mu\nu} - \Lambda g_{\mu\nu} \tag{12.16}$$

One way to look at this is to write

$$G_{\mu\nu} = T'_{\mu\nu} \tag{12.17}$$

so that the cosmological constant just modifies the energy-momentum tensor. In the case of a fluid in equilibrium, we have

$$T'_{\mu\nu} = \begin{pmatrix} \rho - \Lambda & 0 & 0 & 0 \\ 0 & p + \Lambda & 0 & 0 \\ 0 & 0 & p + \Lambda & 0 \\ 0 & 0 & 0 & p + \Lambda \end{pmatrix} \tag{12.18}$$

where we are using the diagonal metric $g_{\mu\nu}$ = (+1, -1, -1, -1). Although this is not the general case, it shows that the role of the cosmological constant is to modify the invariant mass density

$$\rho_{inv} = \left[(\rho - \Lambda)^2 - 3(p + \Lambda)^2 \right]^{1/2} \tag{12.19}$$

which is the source of the gravitational field.

Now, how should we define zero mass/energy? I would think we should define it as the source that gives $G_{\mu\nu} = 0$. If that source happens to be the vacuum, where $T_{\mu\nu} = 0$, we conclude that zero mass/energy is defined so that $\Lambda = 0$.

12.5. ACCELERATION WITH ZERO COSMOLOGICAL CONSTANT

Let us consider the possibility that the cosmological constant is identically zero. In that case, what, then, is responsible for the acceleration of the expansion of the universe? This includes the inflation in the early universe and the current accelerating expansion, both of which have been associated with the cosmological constant, although perhaps not so much for the source of inflation.

I mentioned in chapter 5 that a medium with negative pressure could also produce repulsive gravity. This medium is called *quintessence*.

Let us recall the Friedmann equation (5.2), with $c = 1$,

$$\frac{1}{a}\frac{d^2a}{dt^2} = -\frac{4\pi G}{3}(\rho + 3p) + \frac{\Lambda}{3} \tag{12.20}$$

where ρ is the matter density and p is the radiation pressure. As we showed in chapter 5, we generally take $\rho = p = 0$ at the beginning of the universe. Then the solution is simply

$$a(t) = a(0)\exp\left[\left(\frac{\Lambda}{3}\right)^{\frac{1}{2}} t\right] \tag{12.21}$$

assuming $da/dt = 0$ at $t = 0$. This is exponential inflation.

Suppose instead that $\Lambda = 0$, $p = w\rho$. If ρ = constant

$$a(t) = a(0)\exp\left\{\left[-\frac{4\pi G}{3}\rho(1+3w)\right]^{\frac{1}{2}} t\right\} \tag{12.22}$$

which will yield exponential inflation provided $w < -1/3$. Here ρ is the density of some medium having negative pressure.

So a cosmological constant is not needed for early universe inflation nor for the current cosmic acceleration. Note this is not vacuum energy, which is assumed to be identically zero, so we have no cosmological constant problem and no need for fine-tuning. Note also that the total density need not be a constant.

While current cosmological data are a good fit to a cosmological constant, this does not rule out quintessence. Should a future precise measurement yield a value significantly different, then the quintessence solution would be the alternative.

12.6. THE MULTIVERSE AND THE PRINCIPLE OF MEDIOCRITY

Modern cosmology strongly suggests, although it does not prove, the existence of multiple universes in a greater system called the multiverse.[10] If they exist, multiple universes provide a no-brainer solution to the fine-tuning problem by way of the weak anthropic principle. There are many universes out there with different parameters, and we just happen to be in the one with those parameters that allowed our kind of life to evolve. Our universe is not fine-tuned for life. Life is fine-tuned to our universe.

As we have seen, evidence strongly supports the proposal made independently around 1980 that the universe began with an enormously rapid exponential expansion called inflation. If a process such as this was responsible for our universe, there is no reason it wouldn't produce many others. Further developments of inflationary theories predict just this. Since we want to retain the term *universe* for our familiar cosmos, we call this collection of universes the *multiverse*.

Each universe in the multiverse would be somewhat like our own but not exactly the same. That is, they all would be described by the same set of basic laws of physics, such as energy conservation, that follow from the natural symmetries of space and time. However, they would differ in those properties that follow when other symmetries are spontaneously (accidentally) broken.

Now, many people, scientists as well as educated laypeople, have objected to even discussing multiple universes in a scientific context because they probably can never be detected. However, science talks all the time about undetected, or even undetectable, objects. According to the standard model, individual quarks and gluons are undetectable. Yet they are part of a model that has worked well for three decades. So nothing stops us from considering undetectable universes, as long as they remain consistent with and are suggested by existing well-established theories.

Furthermore, respected cosmologists, such as Alexander Vilenkin and Nobel laureate Steven Weinberg, think that it may be possible to make empirical tests of the multiverse hypothesis. In fact, Vilenkin argues that there already exists some positive evidence.[11]

He bases this in what he calls the *principle of mediocrity*. The idea arises from the statistical fact that when the characteristics of individual objects within a population of similar objects are normally distributed (that is, follow a bell-shaped curve), then when you sample such a population you are more likely to find a property near the average than one out on either tail of the curve. For example, the next adult male you see walking past you on the street is more likely to be in the vicinity of five feet nine inches tall than under five feet or over seven feet.

Applying this principle to the multiverse, we have good reason to believe, as I have outlined, that all universes will obey the same "laws of physics," such as energy conservation, since these are based on symmetry principles that would naturally exist in the absence of any designer. Differences result when symmetries are spontaneously, that is, randomly, broken from one universe to the next. Those properties associated with broken symmetries would then be expected to have a distribution something like, but not necessarily exactly equal to, the normal or Gaussian distribution.

This implies that when we use physics to compute the possible range of a parameter, the value of that parameter in our universe should not be near the edges of that range but somewhere in the mediocre in-between. Of course, one parameter could turn out near the edge by chance, but if we examine a large number of parameters, the multiverse hypothesis would predict an overall grouping in the mediocre region. At least, the hypothesis could be falsified by all the parameters appearing near the edge.

In 1995, Vilenkin published an estimate that the energy density of the cosmological constant should not be much greater then ten times the average density of matter.[12] He also argued that a much smaller value was unlikely. He mentions others, including Weinberg, getting similar results from the same type of "anthropic" reasoning.[13]

In a recent paper posted on the Internet, physicist Stephen Feeney and collaborators have claimed that data on the cosmic microwave background from the WMAP satellite are consistent with the multiverse hypothesis.[14]

Prior to 1998, the common wisdom was that the cosmological constant, Einstein's "greatest blunder," was zero. But then the surprising acceleration of the expansion of the universe was discovered.[15] The dark energy that is deemed responsible for the acceleration is about three times the density of matter. As we saw in chapter 5, the cosmological constant is the favorite candidate for the origin of dark energy, so Vilenkin feels he made a successful prediction. The jury is still out. Although, as we saw above, quintessence is an alternate possibility, Vilenkin argues that this model does not explain why the dark energy density is comparable to the matter density, which is termed the *coincidence problem* in cosmology.

One theist who is perfectly happy with the multiverse is Smithsonian astronomer Owen Gingrich. He reminds us,

"Christians have long envisaged a world in which they have no physical contact, not the heavens but Heaven, the empyrean." This is a place that suspends the rules of our cosmos, which is "tantamount to being in another universe."[16]

I don't read Gingrich as saying that heaven might be one of the universes in the multiverse we have talked about, although apologist Dinesh D'Souza seems to think that's what Gingrich means. D'Souza writes, "If there are multiple universes . . . it is quite conceivable that one of them operates precisely according to the guidelines of the Christian empyrean. . . . Heaven now becomes a real possibility under the existing diversity of laws that govern multiple universes."[17]

If this were the case, heaven would be material in nature and bound (or, its physics would be bound), by a few laws, such as energy conservation. I'm sure Gingrich realizes this and he is just making an analogy.

Cosmologist Don Page, an evangelical Christian, also differs from most other theists in finding the multiverse proposal as supportive of the God hypothesis: "God might prefer a multiverse as the most elegant way to create life and the other purposes He has for His Creation."[18] Of course God, if he exists, could do anything he wants. But why the God of a small, ancient Middle-Eastern tribe? Why not the Force of *Star Wars*?

If scientists can imagine other worlds, why can't believers? The difference is that the scientific belief, and it is a highly tentative belief, is based on well-established physics and cosmology. The religious belief rests on no comparable scientific ground but just the fantasies of the prescientific age. The scenario of the natural origin of the universe that I described above predicts inflation in its equations. It also allows for what I called a *biverse*: our universe along with a mirror universe expanding in the opposite time direction to ours, but with an arrow of time in that direction. If a process such as this were responsible for our universe, there is no reason it wouldn't produce many others.

In an eternal multiverse with an unlimited number of baby universes, one just like ours is likely to occur. But then, why write a book about it? I'm doing so to show that even in the unlikely case that only a single universe exists, there is no fine-tuning, that is, fine-tuning is a fallacy from all angles.

13.
MonkeyGod

13.1. PRINCIPLES AND PARAMETERS

I have argued throughout this book that the models of physics are human contrivances that are designed to describe present observations and predict future ones. In order to be as objective as possible, these models must be formulated in such a way that they do not depend on the point of view or reference frame of the observer. That is, they must possess certain symmetries so that they are invariant to a transformation from one point of view to another. As I mentioned in a previous chapter, I call that principle *point-of-view invariance*.

When the models are so formulated, then according to Noether's theorem they will automatically contain certain conserved quantities, that is, quantities that do not change with time. Mathematically these conserved quantities correspond to the generators of the transformations. Thus, conservation of energy, linear momentum, and angular momentum must necessarily appear in any model that is invariant to translations in time, translations in space, and rotations in space, respectively. Newton's laws of motion follow from momentum conservation, and when you include energy conservation you have the basis of all classical mechanics except for the force laws.

Special relativity similarly results from the principle that the models of physics must be the same for two observers moving at a constant velocity with respect to one another. In general relativity, the gravitational force is treated as a fictitious force like the centrifugal force, introduced into models to preserve invariance between reference frames accelerating with respect to one another.

Generalizing, we found that what is conventionally termed *gauge invariance*, or *gauge symmetry*, is just the application of point-of-view invariance to the abstract space of quantum state vectors. Maxwell's equations and the main properties of quantum mechanics then follow automatically. The standard model of particles and forces that has been so successful for the last thirty years is also based on gauge symmetry. However, in this case gauge symmetry applies fully only at very high energies such as in the early universe and is spontaneously broken in its electroweak sector at the comparatively low energy of current experiments and observations.

Of course, even though the standard model has been around for a generation now, it is only temporary. We can look forward to it being replaced soon, as the data from the Large Hadron Collider (LHC) have started coming in, and theorists will soon have some new observations to anchor their proposed models. Many physicists expect that ultimately a *theory of everything* (TOE) will be discovered that will include a calculation of all its parameters. In that case, there will be nothing to fine-tune.

For my purposes in this book, however, I have stuck to what we already know with some confidence, and I have speculated only very conservatively beyond the standard model when I do it at all. At the same time, I see no reason to try to imagine a universe with different "laws," since there are no such laws that we have access to as humans. Barring revelation, all we know is what we observe, and the best we can do is build models to describe those observations. Unless God exists and breaks point-of-view invariance with his particular point of view, we should be able to describe any and all universes with the same basic models that have symmetries automatically built in and only allow for the possibility of spontaneously broken symmetries. The simplest universe we can imagine and work with is the universe in which we live, even if the universe is a multiverse of many universes.

It happens that the primary properties of our universe that are of interest to most of humanity, from the dimensions of atoms to the lifetime of stars, can be estimated from the values of just four fundamental constants:

α the electromagnetic force strength $k_E e^2 / \hbar c$ at low energy
α_S the strong nuclear force strength at low energy
m_e the mass of the electron
m_p the mass of the proton

The proton mass depends on the masses of quarks, but it can be treated as elementary for this purpose. Recall that the dimensionless gravitational strength is $\alpha_G = Gm_p^2 / \hbar c$, where G, $\hbar$, and c are determined by whatever units we work in. We saw that this is given by m_p. So this does not have to be included as an independent parameter.

It may appear that by using the absolute mass values I have violated the dictum mentioned earlier that only dimensionless parameters are meaningful. However, the units are arbitrary and if I had used Planck units ($\hbar = c = G = 1$), the masses would have been dimensionless and the results the same.

These parameters will not be sufficient to tell a complete story, but they will be sufficient to give the basic properties of a universe and give a feeling for how fine-tuned it really is.

13.2. SIMULATING UNIVERSES

Almost twenty years before this writing, I created a program, MonkeyGod, which can still be executed from my website on the World Wide Web.[1] I first presented results from this program in my 1995 book, *The Unconscious Quantum*.[2] Try your own hand at generating universes! Just choose different values of the four constants and see what happens. While these are really only "toy" universes, the exercise illustrates that many different universes are possible, even within the existing laws of physics. In the following, I present updated results from an improved version of the model used.

As mentioned many times, the constants $\hbar$, c, and G are not considered parameters. They just define the units you choose to use and can all be set to unity with no change in the physics.

I include these constants in the following formulas so that we see the results in more familiar units, although this makes the equations more complicated.

The following are textbook equations:

Bohr radius:

$$r_B = \frac{\hbar}{\alpha m_e c} \tag{13.1}$$

The value in our universe is 5.29×10^{-11} meter.

Ground state binding energy of a hydrogen atom:

$$E_B = \frac{1}{2}\alpha^2 m_e c^2 \tag{13.2}$$

The value in our universe is 13.6 electron-volts.

Radius of a nucleon:

$$r_N = \frac{\hbar}{\alpha_S m_p c} \tag{13.3}$$

The value in our universe is 1.05×10^{-15} meter.

Ground state binding energy of a nucleon:

$$E_N = \frac{1}{2}\alpha^2_S m_p c^2 \tag{13.4}$$

The value in our universe is 18.8 MeV.

An important quantity for our purposes will be the lifetime of main sequence stars, which must be long enough in any universe for life to evolve. Large-scale properties of the universe that depend just on the fundamental parameters are taken from Press and Lightman.[3] These should be understood to be crude estimates.

The maximum mass of a star above which it becomes hydrostatically unstable:

$$M_{star} = \alpha_G^{-3/2} m_p \tag{13.5}$$

The value in our universe is 3.70×10^{30} kg. The mass of our sun is 1.99×10^{30} kg.

The luminosity of a star of mass M:

$$L = 2 \times 10^{-4} \left(\frac{c^5}{G}\right) \alpha_G^5 \alpha^{-2} \left(\frac{m_e}{m_p}\right)^2 \left(\frac{M}{m_p}\right)^3 \tag{13.6}$$

Using $M = M_{star}$, the value in our universe is 3.1×10^{27} Watts.

The maximum lifetime of a star, from (13.6) and (13.5):

$$t_{star} = \frac{M_{star}c^2}{L} \tag{13.7}$$

The value for our universe is 3.39 × 10³ billion years. Note: in my earlier calculations I used a different formula for the lifetime, which I obtained from a paper by Bernard Carr and Martin Rees:[4]

$$t'_{star} = \left(\frac{\alpha^2}{\alpha_G}\right)\left(\frac{m_p}{m_e}\right)^2 \frac{\hbar}{m_p c^2} \tag{13.8}$$

They called that the "minimum lifetime of a main sequence star." Its value in our universe is 0.678 billion years, much lower than 13.7. Since then, I realized it was more reasonable to use the maximum lifetime rather than the minimum lifetime to determine if a universe is capable of supporting life. The program currently on my website still assumes (13.8). The other astronomical parameters follow.

The minimum mass of a planet:

$$M_{\min} = \frac{1}{64} m_p \left(\frac{\alpha}{\alpha_G}\right)^{3/2} \left(\frac{m_e}{m_p}\right)^{3/4} \tag{13.9}$$

The value in our universe is 1.29 × 10²³ kg. The mass of Earth is 5.97 × 10²⁴ kg.

Below this mass, gravity is insufficient for the planet to hold an atmosphere.

Maximum mass of a planet:

$$M_{\max} = m_p \left(\frac{\alpha}{\alpha_G}\right)^{3/2} \tag{13.10}$$

This is the mass above which the object becomes a cold, degenerate dwarf star. The value in our universe is 2.31 × 10^27 kg.

Minimum length of a planetary day:

$$T_{day} = 2\pi(2)^{3/2} \frac{r_B}{c} \left(\frac{m_p}{m_e} \right)^{1/2} \left(\alpha\alpha_G \right)^{-1/2} \quad (13.11)$$

Below this rotation period, a planet will break up because of centrifugal forces. The value in our universe is 5.69 hr.

Maximum year for a habitable planet:

$$T_{year} = 200 \frac{r_B}{c} \left(\frac{m_p}{m_e} \right)^{2} \alpha^{-13/2} \alpha_G^{-1/8} \quad (13.12)$$

where a habitable planet is defined as one in thermal equilibrium at about 350 degrees Kelvin, which corresponds to the characteristic energy of organic bonds. The value in our universe is 17.5.

Finally, the large numbers from chapter 2 are:

$$N_1 = \frac{\alpha}{\alpha_G} \frac{m_p}{m_e} \quad (13.13)$$

The value in our universe is 2.27 × 10^39, and

$$N_2 = \alpha\alpha_S \frac{m_p}{m_e} N_1 \quad (13.14)$$

The value in our universe is 3.04 × 10^39.

None of the astronomical quantities in these approximate formulas depend directly on α_S. The three parameters α, m_e, and m_p alone determine these quantities. However N_2 does depend on α_S.

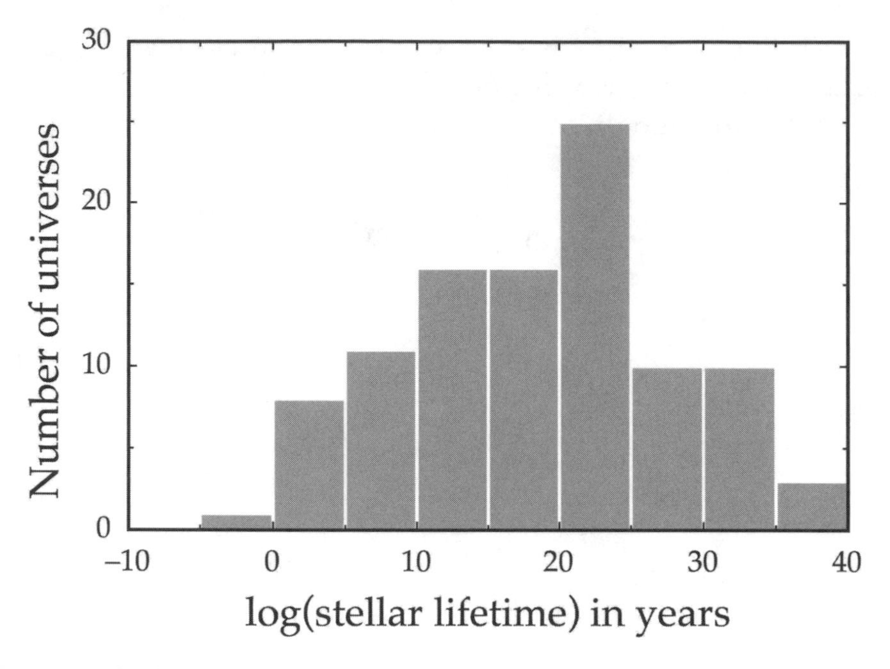

Fig. 13.1. The distribution of maximum stellar lifetimes for 100 random universes where the parameters were varied over ten orders of magnitude.

In previous publications, I have applied MonkeyGod in a computer simulation in which I varied the four parameters randomly (on a logarithmic scale) over ten orders of magnitude. I was mainly interested in seeing how many universes would have stellar lifetimes long enough to allow for some form of life, not necessarily exactly like ours, to evolve.

As mentioned above, I used a formula obtained from one reference for the minimum lifetime of a main sequence star. I later adopted a different formula, from another reference, which I thought was more realistic since it gave the maximum lifetime of a star. Using that formula, but otherwise following the same procedure, I obtained the distribution of maximum stellar lifetimes for 100 universes shown in figure 13.1. The results are not in general disagreement with that previously published.[5] In fact, as expected, I get results even more favor-

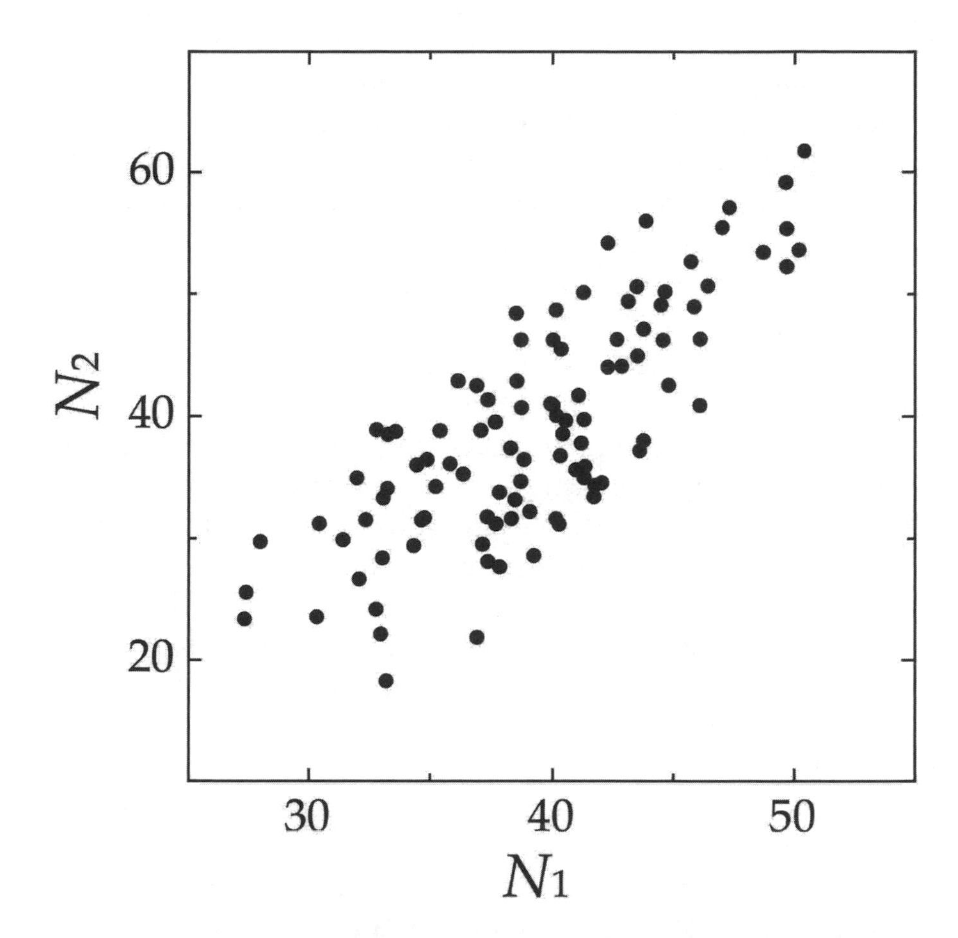

Fig. 13.2. Scatter plot of N_2 versus N_1 for 100 universes in which the values of the four parameters were generated randomly on a logarithmic scale from a range five orders of magnitude below and five orders above their values in our universe.

able to life. While a few stellar lifetimes are low, most are over ten billion years, which is probably enough time for stellar evolution and heavy element nucleosynthesis to occur. Long stellar life may not be the only requirement for biological life, but I have demonstrated that fine-tuning is not necessary to produce a range amenable for life.

Next, let us look at the large number coincidence. Figure 13.2 shows a scatter plot of N_1 and N_2 for the same sample of

100 universes as in figure 13.1. They are not equal in general but are both large numbers and roughly proportional. The N_1–N_2 coincidence, while important historically, is rarely mentioned these days as evidence for fine-tuning.

Christian theologian Robin Collins has raised objections to my preliminary, twenty-year-old conclusion that long stellar lifetimes are not fine-tuned.[6] He argues that not all these universes are livable, that I have not accounted for life-inhibiting features. For example, he says, "if one decreases the strength of the strong nuclear force by more than 50 percent (while keeping the electromagnetic force constant), carbon becomes unstable and with a slightly greater decrease, no atoms with atomic number greater than hydrogen can exist."[7] He refers to Barrow and Tipler, who estimated that $\alpha \leq 11.8\alpha_S$ for carbon to be stable.[8]

Since in this study I was varying all the parameters by ten orders of magnitude, I would not expect such a tight criterion to be satisfied very often. Nevertheless, I have checked and found the Barrow-Tipler limit to be satisfied 59 percent of the time. As we will see below, I have also studied what happens when the parameters are varied by just two orders of magnitude. Then, 91 percent of the time we have $\alpha \leq 11.8\alpha_S$.

Still, I have not taken into account an important fact that is also neglected by the proponents of fine-tuning. As we saw in chapter 10, the standard model of physics and its promising extension, the minimum supersymmetric standard model (MSSM), predict a connection between the force strength parameters. These are expected to be unified, that is, all have the same value, at a very high energy of 3×10^{16} GeV that existed in the early universe. They separate in value as the universe cools, but they are still going to be related. That is, they are not independent variables, and it is unreasonable to expect them to differ by as much as five or ten orders of magnitude at low energies.

Furthermore, recall that not only the force strengths but also the particle masses are constrained by known, well-established physics. Again, we would not expect the masses of the proton and electron to differ by many orders of magnitude.

So a more realistic simulation should be done with parameters limited to a much smaller variation. In the following, I present two sets of results: one set where the parameters are varied by ten orders of magnitude, and one set where they are varied by two orders of magnitude. Both cases are far more than the differences expected in the standard model. In each case, the variation is centered on the values for our universe.

Using a sample of 10,000 universes, I have estimated the fraction of universes generated that should have properties that allow for some form of life not too drastically different from our own. In table 13.1, the set of conditions applied and the fraction of universes that meet those conditions are listed for the two ranges of variation. These conditions are meant to be illustrative, and the reader is cautioned not to take these calculations too literally. Press and Lightman admit that they have neglected certain factors in their formulas that, in some cases, could be off as much as a factor of ten. All I claim here is that there is no evidence for the delicate fine-tuning that its proponents claim.

Other physicists have published their own calculations of the properties of universes with different parameters, and they have arrived at similar conclusions to mine. Fred C. Adams has developed a stellar structure model that enables him to understand the physics of stars with unconventional parameters.[9] He finds that roughly one fourth of the parameter space provides stellar objects that have sustained nuclear fusion. He concludes, "The set of parameters necessary to support stars are not particularly rare."

In chapter 10, I referred to the paper by Roni Harnik and collaborators that shows it was possible to have a viable universe without any weak interactions.[10] However, in this book I am staying within the framework of established physics as given by the standard model.

Table 13.1. List of Conditions Applied to the Universe Simulation and the Fraction That Meet That Condition for Variation of Parameters by (a) Ten and (b) Two Orders of Magnitude

Condition	(a)	(b)
$r_B > 1000\ r_N$	0.61	0.93
$E_B < 1000\ E_N$	0.63	0.96
$\alpha < 11.8\alpha_S$	0.59	0.91
$t_{star} > 10$ billion years	0.61	0.92
$M_{star} > 10\ M_{max}$	0.65	1.00
$M_{max} > 10 M_{min}$	0.84	1.00
$T_{day} > 10$ hours	0.49	0.43
$T_{year} > 100$ days	0.54	0.70
net fraction of livable universes	0.13	0.37

In chapter 11 I also already mentioned the work of Anthony Aguirre, who showed that the cold big-bang alternative to the conventional hot big bang allows for cosmological parameters well outside the region allowed by authors such as Martin Rees.[11] However, as with physics, I can make my case with the standard model of cosmology, the hot big-bang model, alone.

14. Probability

14.1. PROBABILITY ARGUMENTS

We have seen that the question of whether or not a parameter is fine-tuned is a probabilistic one. When Christian apologists Dinesh D'Souza and William Lane Craig quote Stephen Hawking out of context as saying, "if the rate of the universe's expansion one second after the Big Bang had been smaller by even one part in a hundred thousand million million, the universe would have re-collapsed into a hot fireball," they seem to be arguing that the probability for any universe in an ensemble of universes existing with critical density is one part in a hundred thousand million million or, at least, some infinitesimal number. When Hawking follows up a few pages later and informs us that critical density is a consequence of inflationary cosmology, he is saying that the probability for any universe in an ensemble of universes existing with critical density is 100 percent. D'Souza and Craig disingenuously ignore Hawking here.

In the preface I quoted from the 2006 bestselling book *The Language of God: A Scientist Presents Evidence for Belief*, by Francis Collins, one of the most prominent American scientists who is also a devout Christian. Collins was head of the Human Genome Project and at the time of this writing directs the United States National Institutes of Health. In his book, Collins also makes a probability argument. He says there are "fifteen physical constants that the current theory is unable to predict." He includes the speed of light in a vacuum, *c*, which we have seen is arbitrary. He adds:

> The chance that all of these constants would take on the values necessary to result in a stable universe capable of sustaining complex life-forms is almost infinitesimal. And yet those are exactly the parameters that we observe. In sum, our universe is wildly improbable.[1]

Now, a competent scientist such as Collins surely knows that probability arguments are worthless unless they are made comparatively. Collins spends most of his book competently telling us about the DNA basis for life and the processes that go on in the course of evolution. He might have noted that the existence of the species now on Earth, in a term he uses for physics parameters,[2] also rests upon a *knife-edge of improbability*.

Collins does not give us any quantitative estimates to enable us to assess the significance of his "knife-edge of improbability." He might have cited mathematician Roger Penrose's estimate of one part in ten raised to the power of 10^{123} for the probability of the existence of the universe, as we know it, if it had been a random occurrence.[3] If Collins wanted to say something definitive about the need for a creator, he should have compared this with an estimate of the probability that the universe as we know it occurred supernaturally.

Or Collins could have given us some feeling for the significance of the cosmic knife-edge by calculating the probability of his own existence. This, as I have mentioned, is the probability that a particular sperm and egg united multiplied by the probability that his mother and father met, that his grandparents met, and so on back through the generations. I bet he would have obtained an even smaller number. The point is that low-probability events happen every day in this universe. Once they do happen, their probabilities are 100 percent.

14.2. BAYESIAN ARGUMENTS

A statistical method does exist for mathematically estimating how a piece of evidence affects the probability of a statement being true. This method is known as Bayes's theorem after its inventor, Thomas Bayes (d. 1761). Here's how the Bayesian method works: As in symbolic logic, we introduce some symbols that represent logical variables. They can take only two possible values: "true" or "false." The negation of a logical variable V is $\bar{V}$. Let $P(A)$ be the *prior* probability that a proposition A is true. By "prior" probability we mean the probability for a statement being true based on the knowledge that exists before we take into account the new evidence.

Now, suppose we have some new evidence that may be summarized by the proposition E. Let $P(E)$ be the prior probability that E was true before we have made the observation. Let $P(E \mid A)$ be the probability that E is true if A is true. *Bayes's theorem* says that the probability that A is true in light of the new evidence E is

$$P(A \mid E) = \frac{P(E \mid A)P(A)}{P(E)} \tag{14.1}$$

At the time I wrote this I found this nice example in Wikipedia, which I will only slightly paraphrase below.[4]

Suppose there is a school whose students are 60 percent boys and 40 percent girls. The female students wear trousers or skirts in equal numbers; the boys all wear trousers. An observer sees a (random) student from a distance; all the observer can see is that this student is wearing trousers. What is the probability this student is a girl?

We proceed to apply Bayes's theorem. The proposition A is that the student observed is a girl, and the event E is that the student observed is wearing trousers. Let us define some terms:

$P(\mathrm{A})$ is the probability that the student is a girl regardless of any other information. Since the observer sees a random student, meaning that all students have the same probability of being observed, and the fraction of girls among the students is 40 percent, this probability equals 0.4.

$P(\mathrm{E} \mid A)$ is the probability of the student wearing trousers if the student is a girl. As the girls are as likely to wear skirts as they are to wear trousers, this probability is 0.5.

$P(\mathrm{E})$ is the probability of a (randomly selected) student wearing trousers regardless of any other information. Since half the girls and all the boys are wearing trousers, this is $(0.5 \times 0.4) + (1 \times 0.6) = 0.8$.

Given all this information, the probability of the observer having spotted a girl given that the observed student is wearing trousers can be shown to be 0.25.

$$P(\mathrm{A} \mid \mathrm{E}) = \frac{P(E \mid A)P(A)}{P(E)} = \frac{0.5 \times 0.4}{0.8} = 0.25 \qquad (14.2)$$

Getting back to the general situation, since the probability of an event happening plus the probability of it not happening adds up to 1, we can write

$$\begin{aligned} P(E) &= P(\mathrm{E} \mid \mathrm{A})P(\mathrm{A}) + P(\mathrm{E} \mid \bar{\mathrm{A}})P(\bar{\mathrm{A}}) \\ &= P(\mathrm{E} \mid \mathrm{A})P(\mathrm{A}) + P(\mathrm{E} \mid \bar{\mathrm{A}})[1 - P(\mathrm{A})] \end{aligned} \qquad (14.3)$$

where $P(\mathrm{E} \mid \bar{\mathrm{A}})$ is the probability that E is true if A is false. And so,

$$P(\mathrm{A} \mid \mathrm{E}) = \frac{P(\mathrm{E} \mid \mathrm{A})P(\mathrm{A})}{P(\mathrm{A})P(\mathrm{E} \mid \mathrm{A}) + [1 - P(\mathrm{A})]P(\mathrm{E} \mid \bar{\mathrm{A}})} \qquad (14.4)$$

In the 2003 book *The Probability of God: A Simple Calculation That Proves the Ultimate Truth*,[5] physicist Stephen Unwin attempted to calculate the probability that God exists. Unwin's result: 67 percent. Tufts University physicist Larry Ford has examined Unwin's calculation and has made his own estimate using the same formula.[6] Ford's result: 10^{-17} percent.

In what follows, I will compare the two analyses, with some minor modifications that do not change the conclusion.[7]

Let G be the proposition that God exists. Unwin rewrites (0.4) as

$$P_{after} = \frac{P_{before}D}{P_{before}D + 1 - P_{before}} \qquad (14.5)$$

where

$$D = \frac{P(\mathrm{E}\,|\,\mathrm{G})}{P(\mathrm{E}\,|\,\bar{\mathrm{G}})} \qquad (14.6)$$

Unwin defines a "divine indicator," D, which represents how more likely the evidence E would be if God exists compared to him not existing. P_{before} is the probability that God exists *before* the observation of the event E, and P_{after} is the probability that God exists *after* the observation of the event E.

Unwin then puts in some numbers. He takes the prior probability of God existing, that is, the probability before any evidence is submitted, to be $P_{before} = 0.5$. Then he introduces a series of six observations and estimates the divine indicator D for each. At each step he calculates a P_{after} and equates that to P_{before} for the following steps:

1) The evidence for goodness, such as altruism: $D = 10 \Rightarrow P_{after} = 0.91$.
2) The evidence for moral evil done by humans: $D = 0.5 \Rightarrow P_{after} = 0.83$.

3) The evidence for natural evil (natural disasters): $D = 0.1 \Rightarrow P_{after} = 0.33$.
4) The evidence for "intra-natural" miracles (successful prayers, etc.): $D = 2 \Rightarrow P_{after} = 0.5$.
5) The evidence for "extra-natural" miracles (direct intervention by God in nature): $D = 1 \Rightarrow P_{after} = 0.5$.
6) The evidence for religious experience (feeling of awe, etc.): $D = 2 \Rightarrow P_{after} = 0.67$.

Unwin then adds an unjustified, arbitrary boost based on faith, and raises the final probability of God to 0.95.

Now let's look at Ford's alternate estimate of these numbers. First he notes, "Propositions that postulate *existence* have a far less than 50 percent chance of being correct." In other words, the lack of any evidence or other reason to believe some entity such as Bigfoot or the Loch Ness Monster exists implies that it is highly unlikely that it does. So the prior probability of God should be more like one in a million or less. So let's take $P_{before} = 10^{-6}$.

With respect to the divine indicator, D, we must evaluate it for each kind of evidence. Taking miracles, for example, $P(E \mid G)$ is the probability of the observed evidence regarding miracles given that God exists. The observed evidence, E, is that there are no miracles. Since God should be producing them if he existed, this probability is small. On the other hand, the observation that there are no miracles is just what we expect if there is no God, so $P(E \mid \overline{G})$ is near 1. Consequently, the divine indicator based on the absence of evidence for miracles is $D \ll 1$. Let us go though the various indicators.

Unwin exhibits the typical theistic fallacy that goodness can only come from God, and he assigns a high divine indicator, $D = 10$, for this. Ford points out that we should see a lot more goodness in the world than we do see if God exists. So he assumes $D = 0.1$.

Ford notes that the existence of both moral and natural evil in the world is evidence against God's existence. Unwin seems to agree by assigning D values less than 1, but not sufficiently low to describe the true situation in which millions die or suffer needlessly each year from the evils of both humanity and nature. Ford's values of $D = 0.01$ and $D = 0.001$ for moral and natural evil, respectively, are far more reasonable.

Unwin thinks that miracles, such as prayers being answered, have been observed and so assigns a diving indicator $D = 2$ to what he calls "intra-natural" miracles. However, the scientific fact is that the best, controlled experiments on intercessory prayer show no positive effects. These scientific results makes Ford's estimate of $D = 0.01$ in better agreement with the data.

Unwin assigns $D = 1$ for "extra-natural" miracles where God intervenes directly in nature. Since there is not a scintilla of evidence that God does this, including the fact that no miracle was required to bring the universe into existence, Ford's estimate of $D = 0.1$ for this property strikes me as far too generous.

Finally, there is no evidence that so-called religious experiences have any divine content. If they did, we would expect the people having them to return with information about reality that they could not have known before the experience. These "prophecies" could be tested scientifically to see of they came true. None ever have. So, instead of Unwin's $D = 2$, Ford's $D = 0.01$ is also more reasonable.

In any case, here is the summary of Ford's calculation: $P_{before} = 10^{-6}$.

1) The evidence for goodness, such as altruism: $D = 0.1 \Rightarrow P_{after} = 10^{-7}$.
2) The evidence for moral evil done by humans: $D = 0.01 \Rightarrow P_{after} = 10^{-9}$.
(3) The evidence for natural evil (natural disasters): $D = 0.001 \Rightarrow P_{after} = 10^{-13}$.
(4) The evidence for "intra-natural" miracles (successful prayers, etc.): $D = 0.01 \Rightarrow P_{after} = 10^{-14}$.

(5) The evidence for "extra-natural" miracles (direct intervention by God in nature): $D = 0.1 \Rightarrow P_{after} = 10^{-15}$.
(6) The evidence for religious experience (feeling of awe, etc.): $D = 0.01 \Rightarrow P_{after} = 10^{-17}$.

This is to be compared with Unwin's $P_{after} = 0.67$.

Of course, many of you are likely to say this is a silly exercise, that the numbers used are a matter of taste and obvious prejudice. However, I think it is useful to go through it anyway. The mathematically challenged are often awed by any sort of quantitative calculation that they are unable to evaluate and are likely to view Unwin's work as providing scientific support for the existence of God. It does no such thing. Unwin loses. If anything, his method demonstrates the massive unlikelihood of God's existence.

14.3. ANOTHER BAYESIAN ARGUMENT

Bill Jefferys is a retired astronomer from the University of Texas who was the principle investigator for the Hubble Space Telescope Astrometry Science Team, among many other lifetime achievements. He is also one of a number of people I regard as friends with whom I have frequently communicated over the Internet for years but have never met face to face. In addition to astronomy, Jefferys has published many papers on Bayesian statistics and is one of its leading proponents in science. Jefferys and statistician Michael Ikeda have produced a unique Bayesian analysis of the fine-tuning claim.[8] Let me present their argument in some detail because it strikes me as irrefutable.

Ikeda and Jefferys begin with three assumptions:

a) Our universe exists and contains life.
b) Our universe is "life-friendly," that is, the conditions in our universe (such as physical laws, etc.) permit or are compatible with life existing naturalistically.

c) Life cannot exist in a universe that is governed solely by naturalistic law unless that universe is "life-friendly."

While these seem rather obvious, the point is that "a sufficiently powerful supernatural principle or entity (deity) could sustain life in a universe with laws that are not 'life-friendly,' simply by virtue of that entity's will and power."[9] Thus if (1)–(3) are true, then the observation that the universe is "life-friendly" supports the hypothesis that it is governed by naturalistic law. The more finely tuned the universe is, the more the hypothesis of a supernatural creation is undermined.

But these are just words. Let us see the Bayesian analysis of the argument. Let

L = "the universe exists and has life,"
F = "the conditions in the universe are life-friendly," and
N = "the universe is governed solely by natural law."

Using the notation from the previous section, $P(\mathrm{F} \mid \mathrm{N\&L}) = 1$. That is, the probability that F is true given that both N and L are true is 1.

Now the fine-tuning argument says that if $P(\mathrm{F} \mid \mathrm{N}) << 1$, then it follows that $P(\mathrm{N} \mid \mathrm{F}) << 1$. That is, in words, if the probability that a randomly selected universe will be life-friendly given that naturalism is true is very small, then the probability that naturalism is true given the observed fact that the universe is life-friendly is very small. In statistics, this is known as the "prosecutor's fallacy."[10]

Now let's bring in Bayes's theorem. Recall its basic form (14.1). Ikeda and Jefferys define their proposition A = N, that naturalism is true. Their evidence is E = F&L, that is, the universe is life-friendly and life exists. Applying Bayes's theorem,

$$P(\mathrm{N} \mid \mathrm{F} \,\&\, \mathrm{L}) = \frac{P(\mathrm{F} \mid \mathrm{N} \,\&\, \mathrm{L})P(\mathrm{N} \mid \mathrm{L})}{P(\mathrm{F} \mid \mathrm{L})} \tag{14.7}$$

Now, recall proposition (3), which says that in a natural universe life can only exist if that universe is life-friendly. That is, $P(\mathrm{F} \mid \mathrm{N\&L}) = 1$. Thus

$$P(\mathrm{N} \mid \mathrm{F} \,\&\, \mathrm{L}) = \frac{P(\mathrm{N} \mid \mathrm{L})}{P(\mathrm{F} \mid \mathrm{L})} \geq P(\mathrm{N} \mid \mathrm{L}) \tag{14.8}$$

where the inequality follows since probabilities are always less than or equal to 1.

The result of the Bayesian analysis shows that the probability that naturalism is true given that life exists and that the universe is fine-tuned for life is at least as great as the probability that naturalism is true given that life exists whether or not the universe is fine-tuned. Jefferys and Ikeda conclude there is no way that fine-tuning can undermine naturalism.

Note also that

$$\begin{aligned} 1 - P(\mathrm{N} \mid \mathrm{F} \,\&\, \mathrm{L}) &= P(\bar{\mathrm{N}} \mid \mathrm{F} \,\&\, \mathrm{L}) \\ 1 - P(\mathrm{N} \mid \mathrm{L}) &= P(\bar{\mathrm{N}} \mid \mathrm{L}) \\ P(\bar{\mathrm{N}} \mid \mathrm{F} \,\&\, \mathrm{L}) &\leq P(\bar{\mathrm{N}} \mid \mathrm{L}) \end{aligned} \tag{14.9}$$

Let me quote Jefferys's own words in an e-mail message to me:

> Every person knows *a priori* that they exist as sentient beings, that is, that L is true. Anyone who is a sentient being and who

> has an opinion about whether N is true or not can assign priors on N [P(N | L) and P($\bar{N}$ | L)]. Until recently, no one had any idea about F, and indeed, no one had even heard of the notion, which had never been mentioned anywhere until approximately the 1950s. So it was reasonable for anyone to assign such priors, and even today, when hardly anyone knows about F, it [is] reasonable for anyone to assign their priors for those quantities [if] they [have not yet] learned about F (for example, when they [are] young).
>
> The question of interest is: What if, not knowing about the truth or falsity of F, you learn, without question, that F is true? How should this *change your opinion* (that is, what should your posterior look like, compared to your prior)? The answer to this question is the evidentiary value of learning F. Our argument shows that, no matter what you thought before learning F, if you knew that you were a sentient being (L) before you learned that F was true, then learning that F is true can only support your belief in N. That means that F is evidence *for* N, for any sentient being.
>
> Note that this argument is independent of the assignments of P(N | L) and P($\bar{N}$ | L).[11]

I just spent a whole book arguing that the universe is not fine-tuned for life. Suppose I am dead wrong. Ikeda and Jefferys then tell us that the more fine-tuned the universe is for life, the less likely that it was supernaturally created.

Finally, I must mention that University of Wisconsin philosopher Elliott Sober has written extensively on the probability and logical arguments disputing fine-tuning and other design arguments.[12] Links to all his work can be found in his website.[13]

15.

Quantum and Consciousness

15.1. THE NEW SPIRITUALITY

One of the options that John Barrow and Frank Tipler proposed in their 1986 tome *The Anthropic Cosmological Principle* to explain the Strong Anthropic Principle was: *Observers are necessary to bring the universe into being.*[1] They likened this to the philosophy of Bishop George Berkeley (d. 1753) called *idealism,* in which the world is composed of *ideas* whose existence depends on a mind. This does not mean, as often misinterpreted, that reality is all in our heads. There is a real world out there, in this view, but that reality is mental rather than material, whereby "mental" means something composed of some other stuff than matter—what is usually called *spirit*.

While you won't find idealism a common doctrine among philosophers today, the notion that human consciousness can have a profound effect on reality is a growing theme in popular culture. We are currently experiencing a movement in the United States away from traditional religion, especially among young people. A recent survey indicated that 28 percent of

Americans between the ages of 18 and 29 are unaffiliated with any religion.[2] Not all, however, have become atheists or agnostics. About half have adopted what they call "spirituality" based on the instinct that "there has to be something out there" and that something interacts with humanity. What is out there is not the traditional God but a "cosmic consciousness" and the human mind is in tune with that consciousness. Thus we are all one with the universe and it is no coincidence that the parameters of the universe are exactly those that allowed us to be brought into being.

Many have been attracted to ideas that were first developed in the seventies and dubbed the "New Age." Typical New Age thinking can be found in the 1980 book *The Aquarian Conspiracy: Personal and Social Transformation in the 1980s*, by freelance writer Marilyn Ferguson.[3] For a historical and skeptical view, see *Not Necessarily the New Age*, a collection of critical essays edited by Robert Basil.[4]

Physicist Fritjof Capra was a major trigger for the New Age with his 1975 bestseller *The Tao of Physics*.[5] Capra drew analogies between modern physics, particularly quantum mechanics, and Eastern mysticism. In the reductionist view of conventional physics tracing back to the atomic theories of ancient India and Greece and reaffirmed in the nineteenth century, the universe is composed of material objects that can be broken down into parts whose behavior is assumed to be independent of the system as a whole. This agrees with much of common experience. When the water pump in our car breaks down, we can replace the pump with a new one without changing the tires and tuning up the electrical system. Although medical practitioners often pay lip service to "treating the whole person," when you have a bladder infection, they treat that without also giving you a magnetic resonance imaging (MRI) brain scan.

Capra claimed that quantum mechanics and other ideas of modern physics being worked on in the 1970s demonstrated that the whole is greater than the sum of its parts. Thus nature must be understood holistically in much the same way Eastern mystics envisage all reality as one unbroken whole.

However, by the year that *The Tao of Physics* was published (1975), physicists were putting together the standard model of elementary particles and forces. As we have seen, this model reduces all familiar matter to three particles—the up and down quarks and the electron. To account for the less familiar observations physicists make at particle accelerator laboratories and in cosmic rays, the standard model posits a total of thirty-one particles, including antiparticles. These interact by means of two forces, the unified electroweak force and the strong nuclear force. Gravity is treated separately with Einstein's general theory of relativity. The model is totally reductionist, except for some collective effects that occur with systems of identical particles that have been known about since the early days of quantum mechanics.[6]

Although Capra did not make this claim, a primary ingredient in New Age thinking is that quantum mechanics has shown that human consciousness is tuned into reality in very special ways so that we can willfully change the way things are just by thinking about them.

Physicist Amit Goswami explains it this way: "All events are phenomena of consciousness."[7] New Age guru Deepak Chopra made nothing short of revolutionary claims in books published in 1989 and 1993 when he wrote that quantum mechanics shows us that the mind controls reality. In his 2006 book *Life after Death: The Burden of Proof*, Chopra continues that line of argument. He tells us that "a team of innovative scientists" is developing a workable theory of "quantum mind" in which the mind is the controller of the brain.[8] This is contrary to the conventional neuroscientific view, supported by mounds of evidence, that the mind is what the brain does.[9] That is, mind is a manifestation of pure material processes.

In 2009, Chopra published an article in the *San Francisco Examiner* along with doctor of medicine Robert Lanza promoting a "new" theory called *biocentrism*. In the article they assert, "The attempt to explain the nature of the universe, its origins, and what is really going on, including evolution, requires an understanding of how the observer—consciousness—plays a role."[10]

Biocentrism is described in a 2009 book by Lanza and Bob Berman.[11] In the book description, they tell us, "The whole of Western natural philosophy is undergoing a sea change again, forced upon us by the *experimental* findings of quantum *theory* [emphasis added—theories don't have experimental findings]. At the same time, these findings have increased our doubt and uncertainty about traditional physical explanations of the universe's genesis and structure."

Lanza has been making such claims for years. I disputed an article he wrote for the *Humanist*[12] as long ago as 1992.[13] An outstanding detailed review of his book and his more recent article with Chopra has been provided by Vinod K. Wadhawan and Ajita Kamal on the website *Nirmukta* of the Bangalore Skeptics.[14]

15.2. MAKING YOUR OWN REALITY

Quantum consciousness was the theme promoted by Chopra in his 1989 and 1993 books in which he suggested that we can make our own reality, solve all our problems, and heal all our ills by just thinking about them.[15] In recent years, this idea was picked up (and treated as newly discovered) in two highly successful documentary films and books, *What the Bleep Do We Know!?*[16] and *The Secret*.[17]

The book *The Secret*, by Rhonda Byrne and based on the film, tells us how she uncovered The Secret to Life in a one-hundred-year-old book given to her by her daughter Hayley. Tracing it down throughout history, she found that The Secret was known to all the great figures: Plato, Shakespeare, Newton, Hugo, Beethoven, Lincoln, Emerson, Edison, and, of course, Leonardo da Vinci and Einstein. She then searched and found people alive today who know The Secret. The book and the film reveal the profound truths of these remarkable teachers.

Here's how one of these teachers, "philosopher, author, and personal coach" Bob Proctor, understands The Secret:

> The Secret is the law of attraction!
>
> Everything that's coming into your life you are attracting into your life. And it's attracted to you by virtue of the image you're holding in your mind. It's what you are thinking. Whatever is going on in your mind you are attracting to you.[18]

So you can be whatever your want to be just by thinking it. You can be happy, healthy, beautiful, and rich with the right thoughts. "Entrepreneur and moneymaking expert" (appropriate in this context) John Assaraf explains the mechanism:

> What most people don't understand is that a thought has a frequency. We can measure a thought. And so, if you're thinking that thought over and over again, if you're imagining in your mind having that brand new car, having the money you need, building that company, finding your soul mate . . . if you're imagining what that looks like, you're emitting that frequency on a constant basis.[19]

These thoughts are then sent out to the universe and "magnetically attract all *like* things of the same frequency." Proctor tells us, "See yourself living in abundance and you will attract it. It works every time, with every person."[20]

But, then, why isn't everybody living the life of his or her dreams? Assaraf explains: "Here's the problem. Most people are thinking about what they don't want, and they're wondering why it shows up over and over again." Those six million poor Jews killed in the Holocaust. If only they had "thought the right thoughts."

Proctor reveals to us the source of the law of attraction:

"Quantum physicists tell us that the entire Universe emerged from thought! You create your life through your thoughts and the law of attraction."[21]

What do quantum physicists really say?

15.3. WAVES AND PARTICLES

Let me give a brief history of the quantum and see how this idea of quantum consciousness came about.

In 1800, Thomas Young demonstrated that light was a wave phenomenon by observing the interference between beams of light passing through parallel narrow slits in a screen (see figure 15.1, which is discussed below). This phenomenon can be demonstrated with water waves in a pan of water. A hundred years later, Max Planck discovered that light appeared to occur in bundles of energy he called *quanta*.

The minimum energy of each quantum was

$$E = hf \tag{15.1}$$

where f was the frequency of the corresponding electromagnetic wave and h was what is now called Planck's constant, $h = 6.626 \times 10^{-34}$ Joule-seconds.

In 1905, Einstein proposed that these quanta were actually material particles later dubbed *photons*. This proposal successfully explained phenomena, such as the photoelectric effect and atomic spectra, that could not be explained with the wave picture. Thus, light was found to have the seemingly contradictory properties of wave and particle.

In 1925, Louis de Broglie proposed that objects, such as electrons, that are usually identified as localized particles would exhibit wavelike properties, such as double-slit interference.

He proposed that a particle of momentum p had associated with it a wavelength

$$\lambda = \frac{h}{p} \tag{15.2}$$

Note that, for a photon, since its mass is zero, its energy is just its momentum multiplied by c, $E = pc$. The photon's wavelength

$$\lambda = \frac{hc}{E}$$

and, since for waves $f\lambda = c$, then $E = hf$, as in (15.1).

In 1927, Clinton Davisson and Lester Germer verified this prediction experimentally. Eventually, wave properties were observed for other particles, such as neutrons. Thus not only light but also all particlelike phenomena are also wavelike.

Physicists referred to these observations as the *wave-particle duality*. Furthermore, it seemed that which interpretation you gave to an object depended on what you decided to measure. For example, see figure 15.1. There we have Young's classic double-slit experiment with a parallel beam of light incident on two narrow slits. When one slit is covered up, as in (a), we get a localized light-intensity pattern characteristic of a particle. When both slits are open, we do not get two localized patterns, as expected from classical physics, but a light-intensity pattern characteristic of wave interference. So the object seems to be either a particle or a wave depending on the conscious decision of whether or not to mask one of the slits.

Now, as the great physicist John Wheeler pointed out, you could do this experiment with light from a distant galaxy billions of light-years away and delay your choice of what to measure until just before the light entered your apparatus. Then

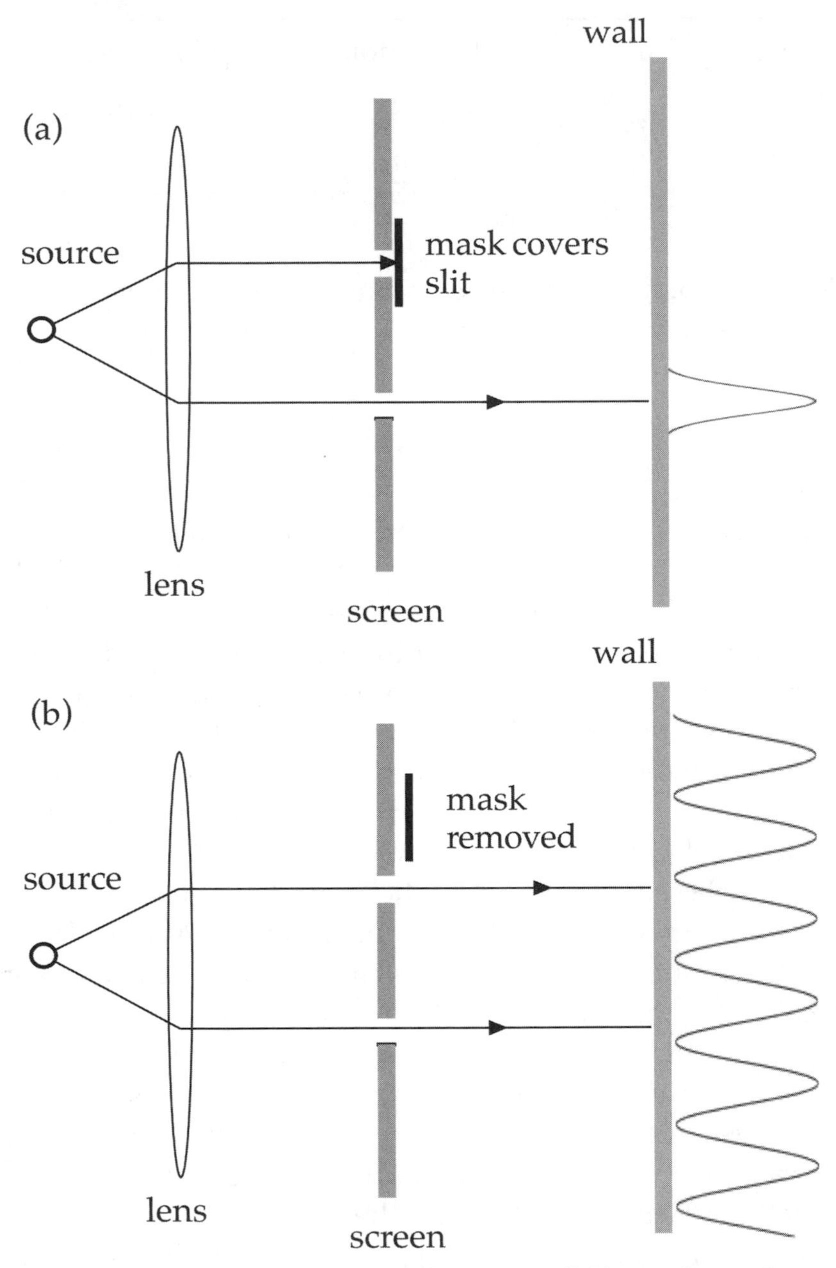

Fig. 15.1. Double-slit experiment. The lens gives a parallel beam of monochromatic light incident on a screen with two narrow slits. In (a), one slit is masked and the light-intensity pattern on the wall is narrow and localized, as expected for a particle. In (b), the mask is removed and the light passing through each slit produces an intensity pattern on the wall characteristic of wave interference.

you would be consciously deciding what the nature of the object is—wave or particle—long after it left its source. This is the profound implication of quantum consciousness that is not usually emphasized by its gurus. The human mind decides the nature of reality, not only here and now, but also more than ten billion years in the past and more than ten billion light-years away—instantaneously. *The Secret* implies that we can change the outcome of historical events just by thinking hard enough. We can make Carthage defeat Rome. We can make Pontius Pilate a forgiving guy. We can have Ralph Branca strike out Bobby Thomson in that last famous playoff game between the Giants and the Dodgers in 1951. Not only that, some sick individual might think the thought that prevents the supernova explosion that would have sprayed carbon and the other heavy chemical elements necessary for planets and life in the vicinity of a certain star, with the result that life never evolves on Earth.

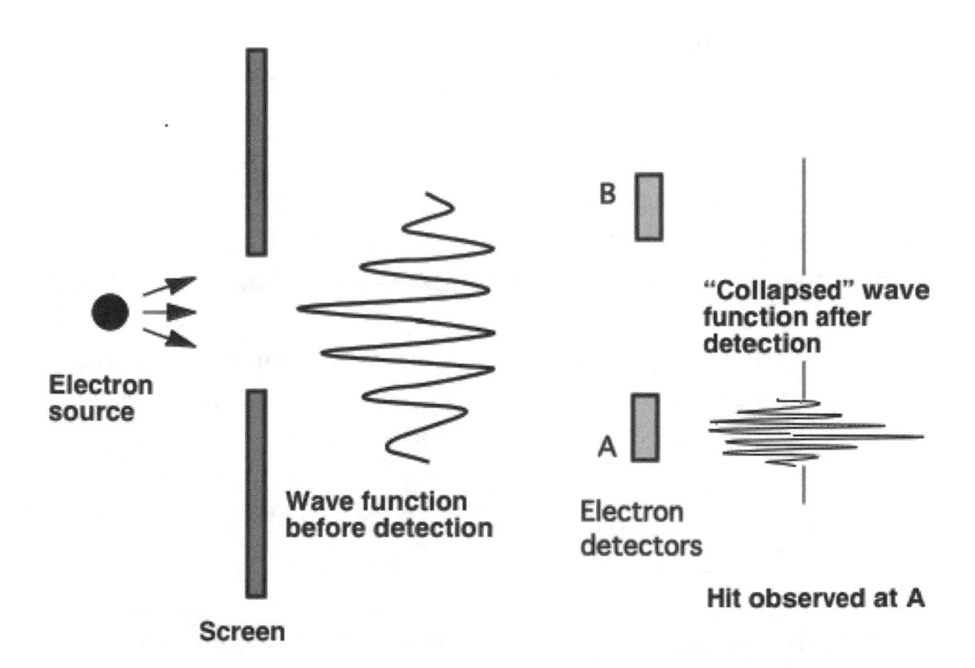

Fig. 15.2. When a particle is detected, its position becomes known to be within the detector. It's state vector, or wave function, is said to "collapse."

However, there is another interpretation that avoids this ridiculous conclusion and still allows us to control the nature of reality. In this interpretation, any given property of a physical system does not become reality until it is measured. Prior to that time, the system has only the potential of giving that experimental result. Since, as I have emphasized, physics observables are defined by how they are measured, then it seems plausible to say that an observable has no meaning until it is measured. Let's see how that works in quantum theory.

As we saw in chapter 4, the state vector $\psi = A\exp(\phi)$ represents the state of a quantum system. The square of the amplitude A gives the probability for finding the system in a unit volume at a particular point in space at a particular time. Since the term *wave function* is probably more familiar, I will use that to describe the state and visually represent it as a "wave-packet," a superposition of sine waves that interfere in such a way that the wave function is zero except in some localized region corresponding to the position of a particle.

Consider the situation described in figure 15.2, which shows the process known as the "collapse of the wave function" that supposedly takes place upon the act of "conscious" measurement. An electron passes though a large hole in a screen on the left, so it is located to the position and area of the hole. Its state is then defined by the wave function, as drawn in the figure just outside the hole.

The particle then heads for two electron detectors A and B. A hit registers in detector A, so the particle is now located more finely, to the position and area of detector A. The wave function is then said to have "collapsed" and is now described by the more narrow wave packet shown just outside A. Since B did not register a hit, the wave function has collapsed to zero in its vicinity.

We can imagine the experiment being done without the initial screen. Then the position of the electron before it is detected is completely unknown. It only becomes known when it is measured.

When my wife and I were in Brussels in June 2009, we stayed at an elegant old hotel, the Metropole. In October 1927, this was the site of the first world conference in the history of physics, and perhaps the greatest, the Fifth *Conseil Solvay*. A photo of the invited-only participants is prominently displayed in the hotel lobby.[22] Seventeen of the twenty-nine invitees were or became Nobel Prize winners. They had gathered to discuss this newfangled theory called "quantum mechanics." There in the photo is Einstein, Schrödinger, Pauli, Heisenberg, Dirac, Madame Curie, Compton, de Broglie, Born, Bohr, and Planck, among others.

It was at the Fifth Solvay Conference that Einstein and Bohr initiated their historic debate on the meaning of quantum mechanics that continued to 1935.[23] At Solvay, Einstein got up to object to the notion of wave function collapse, since it implied an instantaneous action across the universe. Before detection, the electron could have been anywhere in the universe. The moment it is detected, that wave function has collapsed to zero everyplace but within the detector. Twenty years later, Einstein still hadn't changed his mind about this interpretation, writing to Max Born, "Physics should represent a reality in time and space, free from spooky actions at a distance."[24]

The gurus of quantum spirituality are not the only ones who have found that a connection between consciousness and reality is implied by the notions of wave function collapse and wave-particle duality. The great physicist Eugene Wigner said, "The laws of quantum mechanics itself cannot be formulated . . . without recourse to the concept of consciousness."[25] John Wheeler added, "No elementary phenomenon is a phenomenon until it is a registered phenomenon. . . . In some strange sense, this is a participatory universe."[26] Still today you hear authors of popular books quote Wheeler to promote the notion. Physicist Paul Davies and science writer John Gribbin have written, "The observer seems to play an essential role in fixing the nature of reality at the quantum level."[27]

15.4. HUMAN INVENTIONS

When an author says, "a particle has no location until we measure it," this becomes an ontological problem only if we regard the location of a particle as an aspect of objective reality. In my view, the only rational interpretation of the quantities and theories of physics is *instrumental*. Space, time, and the other quantities of physics are human inventions defined *operationally*, by how we measure them. Position is measured with a properly calibrated meter stick. Time is measured with a properly calibrated clock. They are the ingredients of the mathematical models we use to describe observations. These models are used to predict other observations.

Let me give a simple example from high school physics, which should make clear exactly what I am talking about. We drop a rock from a height that we measure with a meter stick to be 15 meters. We have a model that goes back to Galileo that says that the rock will fall with constant acceleration of 9.8 meters per second per second. We use that model to predict that the rock will take 1.7 seconds as measured on a clock to reach the ground.

In no place do we need to know anything about the "true reality" of the position of the rock or the time it hits the ground. We make a measurement with a meter stick and use it to predict a future measurement that we will make with a clock.

So it goes with the quantum "states." They also are simply part of a theory that humans invented to describe what they observe. So when a physicist makes a measurement of a particle's position, she is simply changing the mathematical function that she is using to describe the particle. That function, called the *wave function* for historical reasons, might go from a wide wave packet, which has a poorly defined position, to a sharp "pulse" that now has a well-defined position, as illustrated in figure 15.2.

The vast majority of physicists and neuroscientists give no

credence to the new paradigm of quantum consciousness. Nobel laureate Leon Lederman speaks for most when he calls it all "moo shoo physics." While there are many interesting side issues having to do with holism versus reductionism and with locality versus nonlocality (superluminality), the consciousness claim can be easily discredited without going back to these references.

15.5. A SINGLE REALITY

Unfortunately, the term *wave-particle duality* is confusing without some knowledge of physics or engineering. If you are familiar with the mathematics involved, it is easy to see how the two points of view, wave and particle, are equivalent. They are simply two different ways of saying the same thing.

Engineers have no trouble with the concept because they are familiar with an analogous situation that occurs often in their own work. There a kind of wave-particle dualism also exists in the description of electromagnetic signals that are routinely transmitted and received in radios, televisions, and cell phones. These signals have complex time dependence but are not generally analyzed as a function of time but rather as a function of frequency. This is accomplished by a mathematical procedure called a *Fourier transform*, invented by the great French mathematician Jean Baptiste Joseph Fourier (d. 1830).

The Fourier transform is part of every engineering student's bag of tools. Every time series can be treated as a sum of sine waves of different frequencies, amplitudes, and phases. Given some function of time as in an electromagnetic signal, the student converts it to a function of frequency by way of a Fourier transform. This turns out to be more useful because it determines important parameters such as the *bandwidth* needed to transmit the signal with minimum distortion.

Now, time is a variable used in the description of particle

motion while frequency is a variable used in the description of wave motion. Mathematically it is clear that the two descriptions, wave and particle, are equivalent and describe the same reality.

Conversely, if we start with a temporal signal that is a fairly pure sine wave, the frequency is well determined. Doing a Fourier transform from frequency to time, we get very slow time dependence. That is, when frequency is well defined, the time is ill defined. This illustrated in figure 15.3.

Let us look at a simple example. Suppose we have a signal that is a very narrow pulse in time. This corresponds to a particle passing some point at a well-defined time. If we do a Fourier transform on that signal, we get a broad band of frequencies. That is, when the time is well defined, the frequency is ill defined.

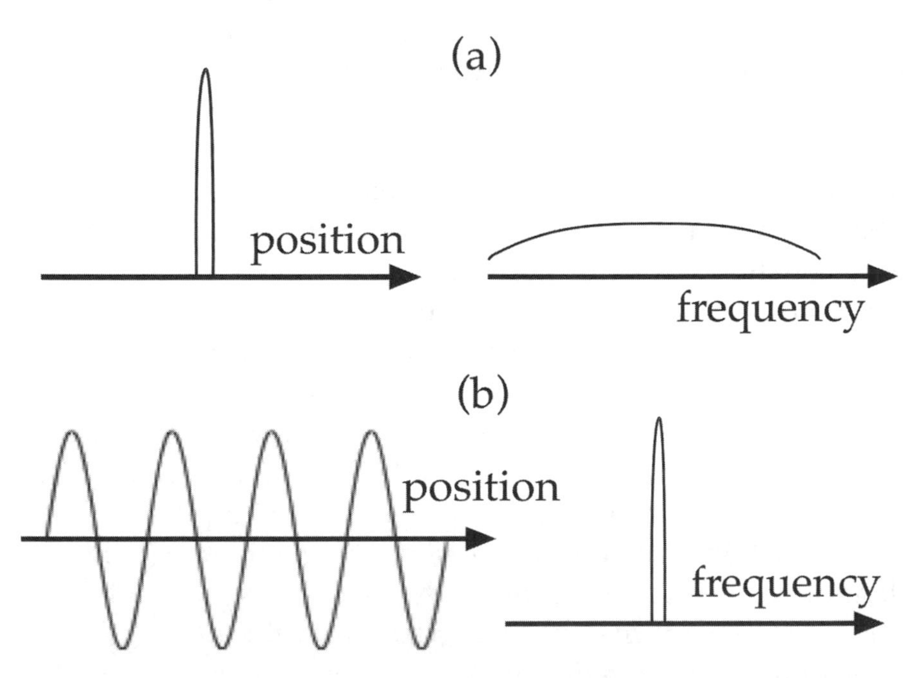

Fig. 15.3. In (a), the position of an object is well determined and its frequency is poorly determined, so we call it a "particle." In (b), the object's frequency is well determined and its position is poorly determined, so we call it a "wave."

Let us look at the mathematics. Suppose $g(t)$ is a function of time. It's Fourier transform is a function of frequency $G(f)$ given by

$$G(f) = \int_{-\infty}^{+\infty} g(t)\exp(-2\pi ift)dt \tag{15.3}$$

Inversely, if we know $G(f)$, then $g(t)$ is given by

$$g(t) = \int_{-\infty}^{+\infty} G(f)\exp(2\pi ift)df \tag{15.4}$$

For example, let us assume we have a pulse in time centered at $t = t_0$, which we can approximate by the Dirac delta-function $\delta(t - t_0)$. Then,

$$G(f) = \int_{-\infty}^{+\infty} \delta(t - t_0)\exp(-2\pi ift)dt = \exp(-2\pi ift_0) \tag{15.5}$$

The *power spectrum*, which gives the energy distribution of the signal among its various frequencies, is $|G(f)|^2 = 1$.

On the other hand, suppose $G(f) = \delta(f - f_0)$, that is, we have a single frequency. Then

$$\begin{aligned} g(t) &= \int_{-\infty}^{+\infty} \delta(f - f_0)\exp(2\pi ift)df = \exp(2\pi if_0t) \\ &= \cos(2\pi if_0t) + i\sin(2\pi if_0t) \end{aligned} \tag{15.6}$$

Although this is a complex number, we see that both the real and the imaginary parts are sinusoids of a fixed frequency. The use of complex numbers in calculations of this sort is common because it usually makes calculations simpler. This was just an easy way of showing that the Fourier transform of a complex sinusoidal wave is a delta function.

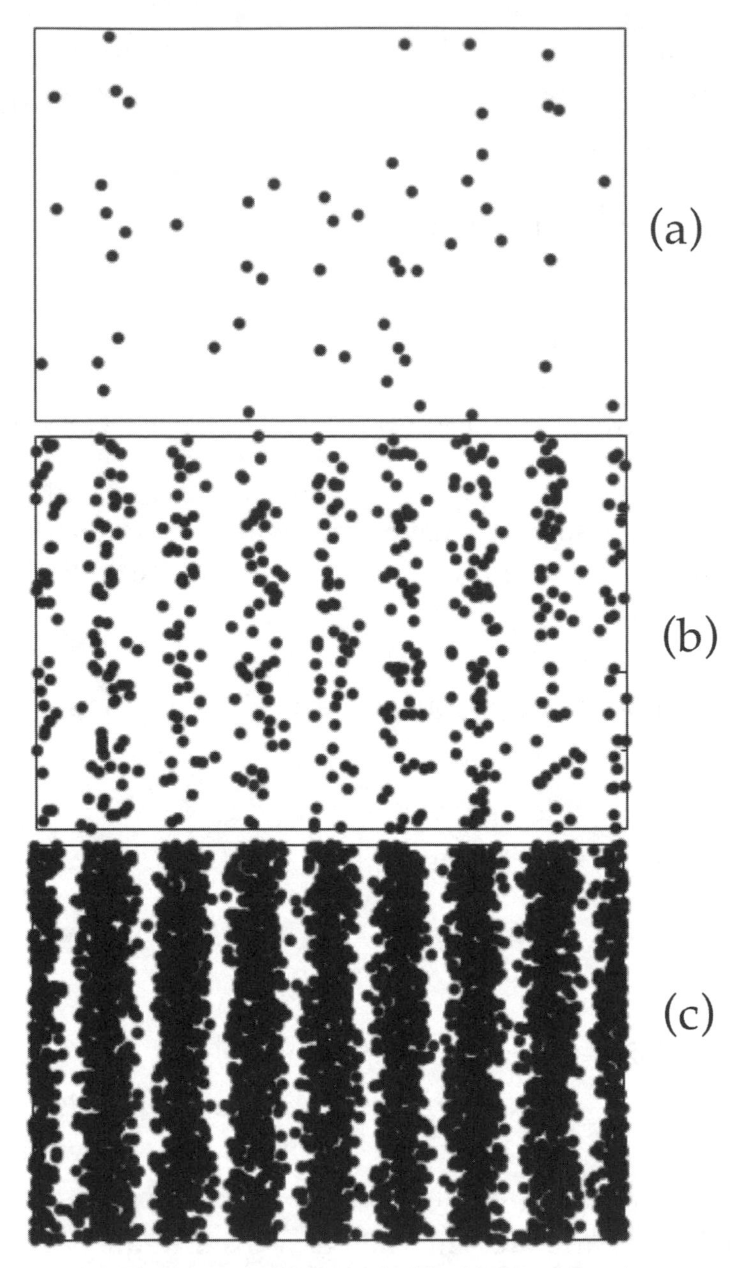

Fig. 15.4. The pattern of hits in a double-slit experiment for an array of small photodetectors sensitive to single photons. In (a), we see the first few hits appear to be randomly distributed. As we accumulate more hits in (b) and (c), the interference pattern associated with wave behavior appears. But it is a statistical effect, not the property of a single particle.

15.6. THE STATISTICAL INTERPRETATION

Fourier transforms are not limited to time and frequency. They work equally well in going from spatial position to wavelength or from wavelength to position. Thus when we "decide to measure" a position, we call the object a "particle." However, we get an ill-defined frequency, so we can't describe it very well as a wave. When we "decide to measure" a frequency, we call the object a wave and get an ill-defined position.

Furthermore, there are experiments that can be done that measure both particles and waves simultaneously. We have seen that the classical way to measure the frequency of a wave is the *double-slit experiment*, which was used by Thomas Young in 1800 to demonstrate the wave nature of light. Today we can do the experiment with a laser source that produces light with a very sharply defined frequency. Then, we can use an array of small, sensitive photodetectors that are capable of registering individual photons. Figure 15.4 shows what the results of such an experiment would look like if we ran it at very low intensity so we could watch the photon hits accumulate one by one. We see that the first few hits seem to be randomly distributed, but, as more hits accumulate, the familiar interference pattern of the double-slit experiment appears. However, that pattern is not associated with a single photon but is the statistical effect of an ensemble of photons.

This empirical result supports the conventional interpretation of the wave function as associated not with individual particles but rather with the probability for finding a particle at a particular position. In this interpretation, the object always is a particle, not a wave, and the wave aspect is a mathematical abstraction used in the model to make probability calculations.

Niels Bohr called wave-particle duality the *principle of complimentarity*. He and many other physicists and philosophers have given this "principle" much more importance than it deserves, resulting in the moo shoo physics we have seen. We see that the idea is quite trivial and provides absolutely no basis for the notion that the human mind can control reality not only here and now but also throughout all time and space.

What physicists always measure can best be modeled as localized particles. As we saw above, the interference and diffraction patterns they observe in sending monochromatic light through narrow slits are not seen in the measurements of individual photons, just in the statistical distributions of an ensemble of many photons. The wave function tells us nothing about an individual particle. It's just an abstract mathematical function that sometimes, not always, is in the functional form of a wave.

15.7. DERIVING THE UNCERTAINTY PRINCIPLE

The Fourier transform can also be used to derive one of the most fundamental ideas in quantum mechanics, the Heisenberg uncertainty principle, which says that the product of the uncertainties in the measurement of a position coordinate and the momentum component along the same axis must be greater than Planck's constant divided by 4π.

When we make a measurement of any quantity, it is never perfect. There will always be some measurement error. If we make a set of measurements, then some distribution of the differences of the measured values from the "true" value will result. That distribution can have any shape in principle, but when the errors are random and the number of measurements is large, it will approach what is called the *normal* or *Gaussian* distribution, the familiar "bell curve." The "uncertainties" involved in the uncertainty principle are the *standard deviations* that measure the width of the statistical distributions.

According to the normal or Gaussian distribution, the probability of obtaining a measurement x within an interval dx is

$$P(x) = \frac{1}{\sigma_x \sqrt{2\pi}} \exp\left[-\frac{(x-\mu)^2}{2\sigma_x^2} \right] \tag{15.7}$$

per unit dx, where $\mu = \langle x \rangle$ is the mean or average value of the measurements of x and $\sigma_x = \langle (x-\mu)^2 \rangle^{\frac{1}{2}}$ is the standard deviation (root-mean square deviation from the mean). The standard deviation is also called the *standard error*, or "uncertainty." It is a measure of the width of the probability distribution.

Now the Gaussian function has the remarkable property that its Fourier transform is also a Gaussian. Thus

$$\begin{aligned} F(k) &= \frac{1}{\sigma_x \sqrt{2\pi}} \int_{-\infty}^{+\infty} \exp\left[-\frac{(x-\mu)^2}{2\sigma_x^2} \right] \exp(-ikx)\, dx \\ &= \sigma_x \sqrt{\frac{2}{\pi}} \exp\left[-2\sigma_x^2 (k-\mu_k)^2 \right] \\ &= \frac{1}{\sqrt{2\pi\sigma_k}} \exp\left[-\frac{(k-\mu_k)^2}{2\sigma_k^2} \right] \end{aligned} \tag{15.8}$$

per unit dk, where $\mu_k = \langle k \rangle$ is the mean value of the measurements of k and $\sigma_k = \langle (k-\mu_k)^2 \rangle^{\frac{1}{2}}$ is the standard deviation in k, where $2\sigma_x \sigma_k = 1$.

Now for a particle, $k = 2\pi / \lambda = 2\pi p / h$, using the de Broglie relation (15.2). So

$$\sigma_x \sigma_p = \frac{h}{4\pi} \tag{15.9}$$

or, if we write the standard errors $\sigma_x = \Delta x$, $\sigma_p = \Delta p$, and $\hbar = h / 2\pi$, we get the familiar form

$$\Delta x \Delta p = \frac{\hbar}{2} \tag{15.10}$$

The equality only holds for a normal distribution of errors. Otherwise this is the minimum product of errors. We see that the more accurately we know one variable, the less accurately we know the other.

16.

Summary and Review

16.1. THE PARAMETERS

Let us now go through all the parameters we have covered and summarize our findings. I am sure you will find it useful to have collected in a single place arguments that are scattered throughout the book. The chapter in which each parameter is discussed in detail is given for reference.

We begin with the five parameters that are supposed to be so fine-tuned that no form of life could exist in a universe in which any of the values differed by an infinitesimal amount from their existing values in our universe. These were listed back in a table in chapter 3, which I repeat here:

Table 16.1. Fine-Tuning of Five Physical Parameters

Parameter	Max. Deviation
Ratio of Electrons to Protons	$1/10^{37}$
Ratio of Electromagnetic Force to Gravity	$1/10^{40}$
Expansion Rate of Universe	$1/10^{55}$
Mass Density of Universe	$1/10^{59}$
Cosmological Constant	$1/10^{120}$

This list is given by Rich Deem on his website.[1] However, I will refer to Deem's major source, Hugh Ross in "Big Bang Model Refined by Fire,"[2] for the cited reasons to believe these are fine-tuned.

Ratio of Electrons to Protons (Chapter 10)

This is Hugh Ross's parameter number 7. He asserts that if it were larger, there would be insufficient chemical binding. If it were smaller, electromagnetism would dominate gravity, preventing galaxy, star, and planet formation.

The number of electrons in the universe should exactly equal the number of protons because of charge conservation, on the reasonable assumptions that the total electric charge of the universe is neutral—as it should be if the universe came from "nothing" and charge is conserved. Charge conservation is one of the principles that follow from point-of-view invariance that should be part of any model that describes any universe unless it is divinely created to have charge or to violate point-of-view invariance.

Conclusion: there is no fine-tuning; the parameter is fixed by established physics and cosmology.

Ratio of Electromagnetic Force to Gravity (Chapters 7 and 13)

This is Ross's number 5. He says that if the ratio were larger, there would be no stars with less than 1.4 solar masses, and hence there would be short and uneven stellar burning. If the ratio were smaller, there would be no stars with more than 0.8 solar masses, and hence there would be no heavy element production.

Note that Ross does not refer to the large number puzzle here, which appears in earlier literature. This is the question of why the force of gravity is 39 orders of magnitude weaker than the electric force in an atom. Recall that this constitutes the earliest of the anthropic coincidences, going back to 1914.

As we saw, the ratio of the forces is calculated for a proton and an electron and depends on their charges and masses. If the masses were much larger, the forces would be closer in value. In fact, the only natural mass in physics is the Planck mass, 2.18×10^{-8} kg. The gravitational force between two Planck mass particles of unit electric charge is actually 137 times stronger than the electric force, so a universe with elementary particles this massive would collapse immediately. Despite the statement often heard in most physics classrooms that gravity is much weaker than electromagnetism, there is no way one can state absolutely the relative strengths of gravity and any other force.

The reason gravity is so much weaker in atoms is that the masses of elementary particles are small relative to the Planck mass. This can be understood to be a consequence of the standard model of elementary particles in which the bare particles all have zero masses and pick up small corrections by their interactions with other particles.

Ross implicitly seems to accept this and argues instead that the ratio of the two forces in atoms must be very close to its existing value in our universe for both stable stars and heavy element production. In my simulation program, MonkeyGod, I varied the strength of electromagnetism α along with the electron mass m_e and the proton mass m_p by both ten and two orders of magnitude centered on their existing values. Note that the units of mass are not important in this analysis since it is their ratios with respect to the Planck mass that matter. Also, recall that the dimensionless strength of gravity is given by $\alpha_G = Gm_p^2 / \hbar c$, so this is not a separate parameter. In disagreement with the claims of fine-tuners everywhere, I find that when the parameters are varied by two orders of magnitude, 37 percent of the universes simulated have the features needed for life similar to ours to evolve, where very strict conditions were applied (see table 13.1).

This example also illustrates a major mistake made by most fine-tuning proponents. They hold all the parameters constant and just vary the one of interest. A proper analysis must vary all parameters at once, since a change on one can often compensate for a change in another. We will see other examples of this below.

Conclusion: there is no fine-tuning; the parameter is in the range expected from established physics.

Expansion Rate of the Universe (Chapter 11)

This is Ross's number 8. He claims that if it were larger, there would be no galaxy formation; if it were smaller, the universe would collapse prior to star formation.

This is also the parameter that apologists William Lane Craig and Dinesh D'Souza referred to when they lifted out of context a quotation from Stephen Hawking's bestseller *A Brief History of Time*, which also bears repeating:

> If the rate of expansion one second after the Big Bang had been smaller by even one part in a hundred thousand million million, the universe would have collapsed before it ever reached its present size.[3]

Craig and D'Souza both ignored the explanation Hawking gave seven pages later for why no fine-tuning was needed:

> The rate of expansion of the universe would automatically become very close to the critical rate determined by the energy density of the universe. This could then explain why the rate of expansion is still so close to the critical rate, without having to assume that the initial rate of expansion of the universe was very carefully chosen.[4]

Conclusion: there is no fine-tuning; the parameter is fixed by established physics and cosmology.

The expansion rate and mean mass/energy density of the universe go hand in hand, so let me bring in the next of Deem's critical parameters.

Mass Density of the Universe (Chapter 11)

While Deem lists this as the "mass of the universe," I am sure he meant mass density. This is Ross's number 10. Ross tells us that if it were larger, there would be too much deuterium from the big bang and stars would burn too rapidly. If it were smaller, there would be insufficient helium from the big bang and too few heavy elements would form.

According to inflationary cosmology, during a tiny fraction of a second after the universe appeared, it expanded exponentially by many orders of magnitude so that it became spatially flat like the surface of a huge balloon. This implied that the mass/energy density of the universe is now very close to its critical value in which the total kinetic energy of all its bodies is exactly balanced by their negative gravitational potential energy. In fact, this was a prediction of inflation that was not an established fact when the model was first proposed. If it had not turned out the way it did, inflation would have been falsified. The success of this prediction is one of several reasons cosmologists consider inflationary cosmology to be a now well-established part of the standard model of cosmology.

The critical density depends on the Hubble parameter, whose inverse is the rate of expansion. The "one part in a hundred thousand million million" that Hawking and the apologists refer to is the precise relation between the density and the Hubble parameter that follows to at least that precision from the inflationary model.

Conclusion: there is no fine-tuning; the parameter is fixed by established physics and cosmology.

The Cosmological Constant (Chapter 12)

Curiously, Ross does not mention this parameter, which is the only one on his list where we still cannot provide a final answer. Deem included it as the fifth parameter in his list in which the value cannot be changed by the tiniest amount without making all forms of life impossible.

Deem gives $1/10^{120}$ as the maximum deviation for the cosmological constant. Apparently he got this number from the result of the calculation, which I repeated in chapter 12, of the total zero-point energy density of boson (integer spin) fields in the universe. Elementary particles are identified as the "quanta" of these fields. The zero-point energy is the energy left over when all the quanta of a field are removed. This calculation comes out 120 orders of magnitude higher than the observed upper limit on the vacuum energy density of the universe.

I pointed out that this calculation ignored the zero-point energy of fermion (half integer spin) fields, which is negative and therefore subtracts from the boson value. If there were an equal number of boson and fermion degrees of freedom, then the total density would be zero. This would be the case if the symmetry between bosons and fermions, called supersymmetry, were perfect. Unfortunately, it is not perfect at low energies, but it is anticipated to be so above 1 TeV. The Large

Hadron Collider (LHC) should confirm or disconfirm if this is so. Supersymmetry reduces the energy density in our cold universe to "only" 10^{50} times the observed value, still a long way from agreement.

The energy density associated with the cosmological constant is the favorite candidate for the dark energy that is presumed to be responsible for the acceleration of the universe's expansion. The dark energy constitutes almost three quarters of the total mass/energy of the universe. Since the universe has an average density equal to the critical density, the dark energy density is almost as big, though still many orders of magnitude below its calculated value. I think it is fair to conclude that the calculation is simply wrong. It should not be taken literally. I will give a good reason why it is wrong in a moment.

Now, physicists still have not reached a consensus on the cosmological question. Some prominent figures such as Steven Weinberg and Leonard Susskind think the answer lies in the anthropic principle applied to multiple universes. In this book I have tried not to depend on the multiverse response to the fine-tuning question, although I disagree with those who say it is unscientific and I do not rule it out. I also have tried to avoid speculations beyond current knowledge.

However, I have mentioned one speculation that is currently getting much attention, the holographic universe. This speculation is based on established physics and so is not simply off the wall.

In calculating the zero-point energy density we summed over all the states in a volume equal to the instantaneous volume of the visible universe. Since the entropy of a system is given by the number of accessible states of the system, the entropy calculated by summing over the volume will be greater than the entropy of a black hole of the same size. But since we cannot see inside a black hole, the information that we have about what is inside is as small as it can be, and so the entropy is as large as it can be. Therefore, it was a mistake to calculate the number of states by summing over the volume. Correcting this by summing over the area, or, equivalently, setting the number of states equal to the entropy of a black hole equal to the size of the universe, we can naturally constrain the vacuum energy density. This says that an empty universe will have a vacuum energy density about equal to the critical density, just the value it appears to have. While there are technical problems with this solution to the dark energy problem, the 50 to 120 order of magnitude calculation is almost certainly wrong.

Another speculation based on existing knowledge is that the so-called "ghost" solutions of relativistic quantum field theory should not be simply dismissed. Including these solutions results in an exact canceling of the vacuum energy density.

Conclusion: The standard calculation of this parameter is grossly wrong and should be ignored. Viable possibilities exist for explaining its value, and until these are all ruled out, no fine-tuning can be claimed.

This takes care of the five parameters that are claimed to be fine-tuned to such precision that even a tiny deviation would make life of any kind impossible. Next let us move to those parameters for which proponents of fine-tuning can only claim that life would be very unlikely if the values of the parameters were different.

The Hoyle Prediction (Chapter 9)

We begin our discussion of the less critical parameters by recapping the story of how in 1951 astronomer Fred Hoyle argued that the carbon nucleus had to have an excited state at 7.7 MeV above its ground state in order for enough carbon to be produced in stars to make life in the universe possible. This story is of great historical interest because it is the only case where anthropic reasoning has led to a seemingly successful prediction. Physicist Lee Smolin asserts, however, that "the anthropic principle (AP) cannot yield any falsifiable predictions, and therefore cannot be part of science."[5]

As we have seen, calculations summarized in figure 9.1 have demonstrated that the same carbon would have been produced if the excited state were anyplace between 7.716 MeV and 7.596 MeV. Sufficient carbon for life would have occurred for an excited state anywhere from just above the ground state to 7.933 MeV. A state somewhere in such a large range is expected from standard nuclear theory. Furthermore, carbon may not be the only element upon which life could be based.

Conclusion: there is no fine-tuning; the parameter is in the range expected from established physics.

Relative Masses of the Elementary Particles (Chapters 8 and 10)

The masses of elementary particles affect many features of the universe, and a number of fine-tuning claims refer to their values. Let me begin with the mass difference between the neutron and the proton. The situation is illustrated in figure 10.1. If the difference in masses between the neutron and the proton were less than the sum of the masses of the electron and the neutrino (the neutrino mass in our universe is negligible for this purpose but may not be in some other universe), there would be no neutron decay. If that were the case, then in the early universe electrons and protons would combine to form neutrons and few, if any, protons would remain. If, on the other hand, the mass difference were greater than the binding energies of nuclei, neutrons inside nuclei would decay, leaving no nuclei behind.

As we see from figure 10.1, there is a range of about 10 MeV for the mass difference to still be in the allowed region for the full periodic table to be formed. The actual mass difference is 1.29 MeV, so there is plenty of room for it to be larger. Since the neutron and proton masses are equal to a first approximation, and the difference results from a small electromagnetic correction, it is unlikely to be as high as 10 MeV.

Next, let us bring in the mass of the electron, which also

affects the neutron decay story. A lower electron mass gives more room in parameter space for neutron decay, while a higher mass leaves less.

The ratio of the electron and proton masses helps determine the region of parameter space for which chemistry is unchanged from our universe, which we saw was quite substantial. No fine-tuning is evident here.

In his parameter number 30, Deem claims fine-tuning for the masses of the three types of neutrinos, saying that if they were smaller, galaxies would not form, and if they were larger, they would be too dense. He is assuming that that the number of neutrinos is a fixed number, so that when the masses are lower, they have less of a gravitational effect, and when the masses are higher, the gravitational effect will be greater. But it is not the total number of neutrinos that is fixed but their total energy. Their gravitational effect depends on the energy and so will be the same regardless of the individual neutrino masses.

Conclusion: there is no fine-tuning; the parameters are in the range expected from established physics.

Relative Strengths of the Forces and Other Physics Parameters (Chapter 10)

The dimensionless relative force strengths are the next set of physical parameters whose fine-tuning is claimed for reasons I found wanting. Let me begin with gravity. We saw that its strength parameter α_G is arbitrary and depends on a mass conventionally taken to be that of the proton. This is a consequence of the fact, already summarized, that there is no absolute way to define the strength of gravity. Thus α_G cannot be fine-tuned. There is nothing to tune.

Next, let me consider the strength of the weak interaction α_W. Ross claims it is fine-tuned to give the right amount of helium and heavy element production in the big bang and in stars. The key is the ratio of neutrons to protons in the early universe when, as the universe cools, their production reactions drop out of equilibrium. As we saw from figure 10.2, a range of parameters is allowed.

What's more, I referred to a paper in which it was shown that universes with no weak interaction might still be able to support life.

The electromagnetic strength is represented by the dimensionless parameter α, historically known as the fine-structure constant, which has the famous value 1/137 at low energies. This is Ross's parameter 4, and he tells us that there would be insufficient chemical bonding if it were different. But, as I showed in chapter 8, the many-electron Schrödinger equation, which governs most of chemistry, scales with α and the mass of the electron. Again, a wide range allows for the chemistry of life.

There are many places where the value of α relative to other parameters comes in. We saw that the weakness of gravity relative to electromagnetism in matter was due to the natural low masses of elementary particles. This can also be achieved with a higher value of α, but it's not likely to be orders of magnitude higher.

The relative values of α and the strong force parameter α_S also are important in several cases. Figure 10.3 shows that when the two are allowed to vary, no fine-tuning is necessary to allow for both nuclear stability and the existence of free protons.

I also pointed out two other facts that most proponents of fine-tuning ignore: (1) the force parameters α, α_S, and α_W are not constant but vary with energy; and (2) they are not independent. The force parameters are expected to be equal at some unification energy. Furthermore, the three are connected in the current standard model and are likely to remain connected in any model that succeeds it. They are unlikely to ever differ by orders of magnitude.

Conclusion: there is no fine-tuning; the parameters are in the range expected from established physics.

Other parameters, such as the decay rate of protons and the

baryon excess in the early universe, have quite a bit of room to vary before they result in excess radiation.

Conclusion: there is no fine-tuning; the parameters are in the range expected from established physics.

Cosmic Parameters (Chapter 11)

We have already disposed of the cosmic parameters that seem so crucial in making any livable universe possible. The mass density of the universe, the expansion rate, and the ratio of the number of protons and electrons are not only not fine-tuned, they are fixed by conventional physics and cosmology.

The deuterium abundance needed for life is small, and a wide range of two orders of magnitude is allowed.

Martin Rees and others have claimed that the lumpiness of matter in the universe, represented by Q, had to be fine-tuned within an order of magnitude to allow for galaxy formation. An order of magnitude is hardly the kind of fine-tuning the theists are claiming. What's more, varying the proton mass along with Q allows, again, for more parameter space for life.

More detailed calculations using the standard concordance model indicate that five parameters contribute to the density fluctuations that provide for galaxy formation. So it is a gross simplification to just talk about varying the single parameter Q. Furthermore, when an alternative—the cold big-bang model—is used, an even wider range of parameters becomes possible.

Conclusion: there is no fine-tuning; the parameter is in the range expected from established physics and cosmology.

Simulating Universes (Chapter 13)

The gross properties of the universe are determined by just three parameters: α, m_p, and m_e. From these we can estimate quantities such as the maximum lifetime of stars, the minimum and maximum masses of planets, the minimum length of a planetary day, and the maximum length of a year for a habitable planet. Generating 10,000 universes in which the parameters are varied randomly on a logarithmic scale over a range of ten orders of magnitude, I find that 61 percent of the universes have stellar lifetimes over 10 billion years, which is sufficient for some kind of life to evolve. Applying rather tight limits to all three parameters, 13 percent of all universes are capable of supporting some kind of life not too different from ours. Varying by two orders of magnitude, which is more realistic since the parameters are not independent but related, I find that 92 percent of the universes have stellar lifetimes over 10 billion years, and 37 percent are capable of supporting some kind of life not too different from ours. Life very different from ours remains possible in a large fraction of the remaining universes, judging from the large stellar lifetimes for most.

Conclusion: there is no fine-tuning; the parameters are in the range expected from established physics and cosmology.

16.2. PROBABILITY ARGUMENTS

The fine-tuning argument is part of a more general set of design arguments for the existence of God that include the origin of the universe in cosmology and intelligent design creationism in biology. The arguments are usually expressed vaguely in the language of probabilities, with no real attention given to the mathematical niceties and difficulties of formal probability theory.

It is claimed that the probability of the set of parameters that describe our universe occurring by natural processes is so incomprehensibly low that they must have been determined by some supernatural process. But nowhere do those who make this claim provide an actual comparison with the probability for the supernatural alternative, without which their statements are worthless. Perhaps that probability is even smaller. After all, no one has ever seen a supernatural event. Natural events with very low a priori probability happen every day, and when they do happen, their probabilities become 100 percent.

In chapter 13, I spent some time applying the Bayesian probability method to the fine-tuning problem and the general problem of God's existence. Basically, the technique is used to estimate the change in probability of a statement being true in the light of new evidence. The absolute value for the statement being true depends on what is called the "prior," that is, the probability that the statement was true before the new evidence was obtained. That prior is usually very difficult to estimate and often ends up controlling the final result. This is especially clear in arguments about the existence of God, for which a believer will bias the calculation one way and an unbeliever, the other.

I gave the example of a calculation by the believing physicist Stephen Unwin, as presented in his 2003 book *The Probability of God: A Simple Calculation That Proves the Ultimate Truth*. Unwin calculates the probability of God to be 67 percent.[6] He begins his calculation by assuming the prior probability of God's existence, before any evidence is considered, to be fifty percent. How does he know that?

I compared Unwin's calculation with the unpublished one of the nonbelieving physicist Larry Ford. Ford used all the same equations as Unwin, differing only in the numbers inserted in the calculation. He argued the prior probability of God must be compared with other unobserved entities such as Big Foot or the Loch Ness Monster, which he took to be one in a million (10^{-6}). With his set of numbers, Ford's conclusion was that the probability God exists is one in a hundred thousand trillion (10^{-17}).

I also applied the Bayesian method to the fine-tuning problem. I followed the procedure of astronomer Bill Jefferys and statistician Michael Ikeda, which concluded that the fine-tuning argument not only does not support the existence of God, it makes a natural universe more probable.[7]

16.3. STATUS OF THE STRONG ANTHROPIC PRINCIPLE

Recall from chapter 2 that Barrow and Tipler listed three options for explaining the strong anthropic principle, which says that the universe must have certain properties to allow the development of life:

1. *There exists one possible universe "designed" with the goal of generating and sustaining "observers."*
2. *Observers are necessary to bring the universe into being.*
3. *An ensemble of other different universes is necessary for the existence of our universe.*

The fine-tuning argument rests on the assumption that any form of life is possible for only a very narrow, improbable range of physical parameters. We can safely conclude that this assumption is completely unjustified. None of this rules out option (1) as the source of the anthropic coincidences. But it does show that the arguments that are used to support that option are very weak and certainly insufficient to rule out all alternatives. If all those alternatives are to fall, making (1) the choice by default, then they will have to fall of their own weight.

Option (2) is the rather solipsistic idea that the universe is all in our heads, "our heads" being the collective consciousness of all humanity. This notion is flimsily based on the careless utterances of many prominent physicists that quantum mechanics introduced the observer into physics and so the properties of physical objects depend in some mysterious way on consciousness. This has encouraged many New Age gurus to claim that we can make our own reality simply by thinking about it.

In chapter 15, I showed that no basis exists for these wild claims. All physical properties depend upon how they are measured. That does not mean that the conscious act of measurement in any way affects the reality of these objects.

Theologian Richard Swinburne has argued that only the multiple-universe scenario of option (3) can explain the anthropic coincidences without a supernatural creator.[8] No doubt that scenario can do it, but I have shown that multiple universes are not necessary to refute fine-tuning. Nevertheless, let me consider for a moment what is called the *multiverse*. If many universes besides our own exist, then they are a simple solution to the problem of the anthropic coincidences. We just happen to live in that universe suited for us. However, most theists scoff at the notion of many universes, saying that they violate Occam's razor.

Actually, the multiverse is more consistent with Occam's razor than the alternative—that only a single universe exists. Occam's razor says that entities are not to be multiplied beyond necessity. Here "entity" is usually interpreted to mean the number of hypotheses in a model, not the number of ingredients. When the atomic model was proposed in the nineteenth century, it multiplied the number of ingredients in a gram of matter by 10^{24}. Yet there can be no doubt the atomic model was simpler, had fewer hypotheses, and was a vast improvement over previous ideas.

Similarly, the multiverse has fewer hypotheses than the single universe model. We have no reason to think there is only one universe and, indeed, modern cosmology suggests there are many. The single universe model thus requires the additional hypotheses that only one universe exists, and also requires some new principle to justify that assumption, violating Occam's razor.

Nevertheless, in this book I have avoided relying on multiple universes. Even if only one universe exists, my analysis in previous chapters shows that the likelihood of some form of life in that single universe is not provably negligible, which is the basic claim of those who embrace anthropic reasoning.

16.4. A FINAL CONCLUSION

We have seen that the proponents of fine-tuning make serious errors in physics, cosmology, probability theory, and data analysis. While not every proponent makes every error, there is a remarkable similarity to the arguments you find in the theist literature, which leads you to suspect that they do not do much research to support their claims beyond simply reading each other's books. Let me list the errors of proponents that I have pointed out in the course of this book:

1. They make fine-tuning claims based on the parameters of our universe and our form of life, ignoring the possibility of other life-forms.
2. They claim fine-tuning for physics constants, such a c, $\hbar$, and G, whose values are arbitrary.
3. They assert fine-tuning for quantities, such as the ratio of electrons to protons, the expansion rate of the universe, and the mass density of the universe, whose values are precisely set by cosmological physics.
4. They assert that the relative strengths of the electromagnetic and gravitational forces are fine-tuned, when, in fact, this quantity cannot be universally defined.
5. They assert that an excited state of the carbon nucleus had to be fine-tuned for stars to produce the carbon needed of life, when calculations show a wide range of values for the energy level of that state.
6. They claim fine-tuning for the masses of elementary particles, when the ranges of these masses are set by well-established physics and are sufficiently constrained to give some form of life.
7. They assume the strengths of the various forces are constants that can independently change from universe to universe. In fact, they vary with energy, and their relative values and energy dependences are close to being

pinned down by theory, in ranges that make some kind of life possible.

8. They make a serious analytical mistake in always taking all the parameters in the universe to be fixed and varying only one at a time. This fails to account for the fact that a change in one parameter can be compensated for by a change in another, opening up more parameter space for a viable universe.
9. They misunderstand and misuse probability theory.
10. They claim many parameters of Earth and the solar system are fine-tuned for life, failing to consider that with the hundreds of billions of planets that likely exist in the visible universe, and the countless number beyond our horizon, a planet with the properties needed for life is likely to occur many times.

I also think the fine-tuners are wrong to reject the multiverse solution as "unscientific." I will not call that an error, since the issue is arguable, whereas the ten examples above are not. But I do not think it is unscientific to speculate about invisible, unconfirmed phenomena that are predicted by existing models that, so far, agree with all the data. The neutrino was predicted to exist in 1930, based on the well-established principle of energy conservation, and was not detected until 1956—and even then indirectly. If the physics community used the fine-tuners' criterion, my late colleague Frederick Reines and his collaborator Clyde Cowen would not have been able to get the money and the support to engage in their search for the neutrino.

With so many errors and misjudgments, and with such a gross lack of understanding of the basic science we have seen exhibited by the supporters of supernatural fine-tuning, we can safely say that their motivation is more wishful thinking than truthful scientific inference. A proper analysis finds there is no evidence that the universe is fine-tuned for us.

Finally, I should mention that knowledgeable theists themselves are beginning to cast doubt on the proposition that fine-tuning is evidence for God. In a January 2011 Internet paper by the prominent cosmologist Don Page, who was a student of Stephen Hawking and who also happens to be an evangelical Christian. Page points out that the apparent positive value of the cosmological constant is somewhat inimical to life since its repulsion acts against the gravitational attraction needed to form galaxies. If God fine-tuned the universe for life, he would have made the cosmological constant slightly negative.[9]

Page joins many theists in simply saying that God could have made the universe any way he wanted. Of course this is true, but Page has realized that theists can no longer claim that the universe is so fine-tuned for humans that the existence of a creator is an inevitable conclusion from the data.

Notes

PREFACE

1. The tale I tell here was one I heard as a child and is unlikely to be completely accurate. However, even if partially fictional, I hope you will find it a good story that makes an important point.

2. Roger Penrose, *The Emperor's New Mind: Concerning Computers, Minds, and the Laws of Physics* (Oxford; New York: Oxford University Press, 1989), pp. 342–44.

3. Edward Robert Harrison, *Masks of the Universe* (New York; London: Macmillan; Collier Macmillan, 1985), p. 252.

4. Francis Collins, *The Language of God: A Scientist Presents Evidence for Belief* (New York: Free Press, 2006), p. 75.

5. Michael Anthony Corey, *God and the New Cosmology: The Anthropic Design Argument* (Lanham, MD: Rowman & Littlefield, 1993), pp. 44–45.

6. George Greenstein, *The Symbiotic Universe: Life and Mind in the Cosmos* (New York: Morrow, 1988), p. 27.

7. Tony Rothman, "A 'What You See Is What You Beget' Theory," *Discover*, May 1987, 99.

8. William Lane Craig, "Debates," http://www.leaderu.com/offices/billcraig/menus/debates.html (accessed February 12, 2010).

9. William Lane Craig, "The Craig-Pigliucci Debate: Does God

Exist?" http://www.leaderu.com/offices/billcraig/docs/craig-pigliucci0.html (accessed February 13, 2010).

10. Stephen Hawking, *A Brief History of Time: From the Big Bang to Black Holes* (New York: Bantam, 1988), pp. 121–22.

11. Paul Davies, *Other Worlds* (London: Dent, 1980), p. 6.

12. John Barrow and Frank Tipler, *The Anthropic Cosmological Principle* (Oxford; New York: Oxford University Press, 1986).

13. Paul Davies, *The Mind of God: The Scientific Basis for a Rational World* (New York: Simon & Schuster, 1992), p. 16.

14. Robert Jastrow, *God and the Astronomers*, 2nd ed. (New York: Norton, 1992), p. 107.

15. Victor J. Stenger, *Has Science Found God? The Latest Results in the Search for Purpose in the Universe* (Amherst, NY: Prometheus Books, 2003); *God: The Failed Hypothesis—How Science Shows That God Does Not Exist* (Amherst, NY: Prometheus Books, 2007); *The New Atheism: Taking a Stand for Science and Reason* (Amherst, NY: Prometheus Books, 2009).

16. Leonard Susskind, *Cosmic Landscape: String Theory and the Illusion of Intelligent Design* (Boston: Little, Brown, 2005).

17. I use the term *atoms* here in the original sense from ancient Greece as the uncuttable objects that make up the universe. They are not to be confused with the chemical elements, which are also called "atoms" for historical reasons but can be subdivided into more elementary ingredients.

18. Robin Collins, "The Teleological Argument: An Exploration of the Fine-Tuning of the Universe," in *The Blackwell Companion to Natural Theology*, ed. William Lane Craig and James Porter Moreland (Chichester, UK; Malden, MA: Wiley-Blackwell, 2009), 202–81.

19. Michael Martin and Ricki Monnier, *The Impossibility of God* (Amherst, NY: Prometheus Books, 2003).

CHAPTER 1: SCIENCE AND GOD

1. Victor J. Stenger, *God: The Failed Hypothesis—How Science Shows That God Does Not Exist* (Amherst, NY: Prometheus Books, 2007).

2. Stephen Jay Gould, *Rocks of Ages: Science and Religion in the Fullness of Life* (New York: Ballantine, 1999).

3. The original is supposedly, "The Bible shows the way to go to heaven, not the way the heavens go."

4. For the latest arguments in natural theology, see William Lane Craig and James Porter Moreland, eds., *The Blackwell Companion to Natural Theology* (Chichester, UK; Malden, MA: Wiley-Blackwell, 2009).

5. William Paley, *Natural Theology; or, Evidences of the Existence and Attributes of the Deity* (London: Wilks and Taylor, 1802).

6. David Hume, *Dialogues concerning Natural Religion*, ed. J. M. Bell (London; New York: Penguin Books, 1990). First published in 1776.

7. David Hume, *An Enquiry concerning Human Understanding*, ed. P. F. Millican (Oxford; New York: Oxford University Press, 2007). First published in 1748.

8. Charles Darwin, *On the Origin of Species: By Means of Natural Selection, or the Preservation of Favoured Races in the Struggle for Life*, ed. David Quammen (New York: Sterling, 2008). First published in 1859.

9. For excellent recent surveys, see Jerry Coyne, *Why Evolution Is True* (Oxford; New York: Oxford University Press, 2009) and Richard Dawkins, *The Greatest Show on Earth: The Evidence for Evolution* (New York: Free Press, 2009).

10. See Barbara Forrest and Paul Gross, *Creationism's Trojan Horse: The Wedge of Intelligent Design* (Oxford; New York: Oxford University Press, 2004).

11. Frank Newport, "Republicans, Democrats Differ on Creationism," Gallup, posted June 20, 2008, http://www.gallup.com/poll/108226/Republicans-Democrats-Differ-Creationism.aspx (accessed July 20, 2010).

12. Vision Critical, "Americans Are Creationists; Britons and Canadians Side with Evolution," posted July 15, 2010, http://www.visioncritical.com/2010/07/americans-are-creationists-britons-and-canadians-side-with-evolution/ (accessed July 20, 2010).

13. Pope John Paul II, "Magisterium Is Concerned with Question of Evolution, for It Involves Conception of Man," posted October 22, 1996, last modified December 3, 2010, http://www.cin.org/jp2evolu.html (accessed July 12, 2009).

14. Some authors call evolution by accident and natural selection a form of "design," but I consider that misleading. It is not only not intelligent design, it is not even dumb design.

15. Michael Behe, *Darwin's Black Box: The Biochemical Challenge to Evolution* (New York: Free Press, 1996).

16. William Dembski, *Intelligent Design: The Bridge between Science & Theology* (Downers Grove, IL: InterVarsity Press, 1999).

17. Dembski claimed that he was making no identification of the intelligent designer with a supernatural entity, which could just as well be an advanced alien civilization. But such a civilization would presumably be natural and so this would violate the law of conservation of information. Furthermore, Dembski's subtitle makes it clear that he is thinking of a supernatural creator.

18. Victor J. Stenger, *Has Science Found God? The Latest Results in the Search for Purpose in the Universe* (Amherst, NY: Prometheus Books, 2003).

19. Claude Shannon and Warren Weaver, *The Mathematical Theory of Communication* (Urbana: University of Illinois Press, 1998).

20. John Jones III, "Kitzmiller v. Dover Area School District et al., Case No. 04cv2688," Filed December 20, 2005, http://www.pamd.uscourts.gov/kitzmiller/kitzmiller_342.pdf, 2005 (accessed January 5, 2011).

CHAPTER 2: THE ANTHROPIC PRINCIPLES

1. Hermann Weyl, *Annals of Physics* 59 (1919): 102.

2. Arthur Eddington, *The Mathematical Theory of Relativity* (London: Cambridge University Press, 1923), p. 167.

3. The cosmic horizon is defined by the distance from us beyond which light cannot reach us in the age of the universe.

4. Charles Piazzi Smyth, *The Great Pyramid: Its Secrets and Mysteries Revealed* (New York: Bell Publishing, 1978).

5. William Steibing, *Ancient Astronauts, Cosmic Collisions, and Other Popular Theories about Man's Past* (Amherst, NY: Prometheus Books, 1984), pp. 108–10.

6. Paul Dirac, "The Cosmological Constants," *Nature* 139 (1937): 323.

7. Robert Dicke, "Dirac's Cosmology and Mach's Principle," *Nature* 192 (1961): 440.

8. Edwin Salpeter, "Accretion of Interstellar Matter by Massive Objects," *Astrophysical Journal* 140 (1964): 796–800.

9. John Barrow and Frank Tipler, *The Anthropic Cosmological Principle* (Oxford; New York: Oxford University Press, 1986), p. 252.

10. Fred Hoyle, "The Universe: Past and Present Reflections," *Engineering and Science* (November 1981): 12.

11. Brandon Carter, "Large Number Coincidences and the Anthropic Principle in Cosmology," in *Confrontation of Cosmological Theory with Astronomical Data*, ed. M. S. Longair (Derdrecht: Reidel, 1974), 291–98.

12. Don Page, "Preliminary Inconclusive Hint of Evidence against Optimal Fine-Tuning of the Cosmological Constant for Maximizing the Fraction of Baryons Becoming Life," Cornell University Library, last modified January 28, 2011, http://xxx.lanl.gov/abs/1101.2444 (accessed January 20, 2011); Don Page, "Does God So Love the Multiverse," in *Science and Religion in Dialogue*, vol. 1, ed. M. Y. Stewart (Chichester, UK: 2010), 296–410.

13. Barrow and Tipler, *Anthropic Cosmological Principle*.

14. Ibid., p. 16.

15. Ibid., pp. 21–23.

16. Martin Gardner, "Wap, Sap, Pap, and Fap," *New York Review of Books* 23, no. 8 (May 8, 1986): 22–25.

17. Rich Deem, "Evidence for the Fine-Tuning of the Universe," God and Science, http://www.godandscience.org/apologetics/designun.html (accessed November 27, 2008).

18. Hugh Ross, *The Creator and the Cosmos: How the Greatest Scientific Discoveries of the Century Reveal God* (Colorado Springs, CO: NavPress, 1995).

19. Reasons to Believe, http://www.reasons.org/ (accessed December 20, 2008). Italics original.

20. Reasons to Believe, "Our Beliefs," http://www.reasons.org/resources/apologetics/index.shtml#design_in_the_universe (accessed December 20, 2008).

21. Hugh Ross, "Big Bang Model Refined by Fire," in *Mere Creation: Science, Faith & Intelligent Design*, ed. William Dembski (Downers Grove, IL: InterVarsity Press, 1988), 363–83.

22. David Reuben Stone, *Atheism Is False: Richard Dawkins and the Improbability of God Delusion* (Raleigh, NC: Lulu, 2007).

23. Dean Overman, *A Case against Accident and Self-Organization* (New York: Rowman & Littlefield, 1997).

24. Dinesh D'Souza, *Life after Death: The Evidence* (Washington, DC: Regenery, 2009).

25. Francis Collins, *The Language of God: A Scientist Presents Evidence for Belief* (New York: Free Press, 2006).

26. John Barrow, *The Constants of Nature: From Alpha to Omega* (London: Jonathan Cape, 2002).

27. Martin Rees, *Just Six Numbers: The Deep Forces That Shape the Universe* (New York: Basic Books, 2000). First published in 1999.

28. Paul Davies, *The Goldilocks Enigma: Why Is the Universe Just Right for Life* (London: Allen Lane, 2006).

29. Richard Swinburne, "Argument from the Fine-Tuning of the Universe," in *Modern Cosmology & Philosophy*, ed. John Leslie (Amherst, NY: Prometheus Books, 1999), 160–79.

30. John Leslie, "The Anthropic Principle Today," in *Modern Cosmology & Philosophy*, ed. Leslie (see note 28), 289–310.

31. Leslie, *Modern Cosmology & Philosophy* (see note 28).

CHAPTER 3: THE FOUR DIMENSIONS

1. Susan Haack, *Defending Science—Within Reason: Between Scientism and Cynicism* (Amherst, NY: Prometheus Books, 2003), pp. 33–43.

2. There have been attempts to formulate physics models without space and time, notably the *bootstrap theory* of the 1960s. These have not proved useful.

3. Martin Rees, *Just Six Numbers: The Deep Forces That Shape the Universe* (New York: Basic Books, 2000), pp. 149–63. First published in 1999.

4. Relativity allows for the theoretical possibility of particles, called *tachyons*, that always travel at the speed of light and cannot be decelerated below that limit. None have been observed.

5. Rich Deem, "Evidence for the Fine-Tuning of the Universe," http://www.godandscience.org/apologetics/designun.html (accessed November 27, 2008).

6. These uncertainties are equal to the standard error as defined in statistics.

7. Deem, "Evidence for the Fine-Tuning of the Universe."

8. Let us not take the number thirty-seven too literally. The number of adjustable parameters in the standard model is not set in concrete.

9. Hugh Ross, *The Creator and the Cosmos: How the Greatest Scientific Discoveries of the Century Reveal God* (Colorado Springs, CO: NavPress, 1995), pp. 138–41.

CHAPTER 4: POINT-OF-VIEW INVARIANCE

1. Nina Byers, "Emmy Noether," in *Out of the Shadows: Contributions of Twentieth-Century Women to Physics*, ed. Nina Byers and Gary Williams (Cambridge, UK; New York: Cambridge University Press, 2006).

2. Robin Collins, "The Teleological Argument: An Exploration of the Fine-Tuning of the Universe," in *The Blackwell Companion to Natural Theology*, ed. William Lane Craig and James Porter Moreland (Chichester, UK; Malden, MA: Wiley-Blackwell, 2009), 212.

3. Paul Dirac, *The Principles of Quantum Mechanics* (Oxford: Clarendon Press, 1930). This book has had four editions and at least twelve separate printings.

4. Ibid., p. vii.

5. Ibid., p. 80.

6. Freeman Dyson, "Feynman's Proof of Maxwell's Equations," *American Journal of Physics* 58, no. 3 (1989): 543–44.

7. Victor J. Stenger, *Timeless Reality: Symmetry, Simplicity, and Multiple Universes* (Amherst, NY: Prometheus Books, 2000).

CHAPTER 5: COSMOS

1. Max Tegmark et al., "Towards a Refined Cosmic Concordance Model: Joint 11-Parameter Constraints from the Cosmic Microwave Background and Large-Scale Structure," *Physical Review D* 63 (2000): 043007.

2. Laurent Nottale, "Scale Relativity and Fractal Space-Time: Theory and Applications," *Foundations of Science* 15, no. 2 (2008): 101–52.

3. Demos Kazanas, "Dynamics of the Universe and Spontaneous Symmetry Breaking," *Astrophysical Journal* 241 (1980): L59–L65.

4. Alan Guth, "Inflationary Universe: A Possible Solution to the Horizon and Flatness Problems," *Physical Review D* 23, no. 2 (1981): 347–56.

5. Andrei Linde, "A New Inflationary Universe Scenario: A Possible Solution of the Horizon, Flatness, Homogeneity, Isotropy and Primordial Monopole Problems," *Physics Letters B* 108 (1982): 389.

6. Alan Guth, *The Inflationary Universe: The Quest for a New Theory of Cosmic Origins* (Reading, MA: Addison-Wesley Publishing, 1997).

7. Ibid., p. 185.

8. G. Smoot and K. Davidson, *Wrinkles in Time* (New York: Morrow, 1993).

9. E. Komatsu et al., "WMAP Five-Year Observations: Cosmological Interpretation," *Astrophysical Journal Supplement Series* 180 (2009): 330–76.

10. Jacob D. Bekenstein, "Black Holes and Entropy," *Physical Review D* 7, no. 8 (1973): 2333–46.

CHAPTER 6: THE ETERNAL UNIVERSE

1. William Lane Craig, *The Kalâm Cosmological Argument. Library of Philosophy and Religion* (London: Macmillan, 1979).

2. Roy Jackson, *The God of Philosophy: An Introduction to the Philosophy of Religion* (London: Philosophers' Magazine, 2001), pp. 21–37.

3. William Lane Craig and James D. Sinclair, "The *Kalâm* Cosmological Argument," in *The Blackwell Companion to Natural Theology*, ed. William Lane Craig and James Porter Moreland (Chichester, UK; Malden, MA: Wiley-Blackwell, 2009), pp. 101–201.

4. Ibid., p. 102.

5. David Hume, *An Enquiry concerning Human Understanding* (Chicago: Gateway Editions, 1956).

6. David Bohm and B. J. Hiley, *The Undivided Universe: An Onto-*

logical Interpretation of Quantum Theory (London; New York: Routledge, 1993).

7. Victor J. Stenger, *The Unconscious Quantum: Metaphysics in Modern Physics and Cosmology* (Amherst, NY: Prometheus Books, 1995).

8. David Hilbert, "On the Infinite," in *Philosophy of Mathematics*, ed. Paul Benacerraf and Hillary Putnam (Englewood Cliffs, NJ: Prentice-Hall, 1964), 139–41.

9. William Lane Craig, "The Craig-Pigliucci Debate: Does God Exist?" http://www.leaderu.com/offices/billcraig/docs/craig-pigliucci0.html (accessed February 13, 2010).

10. Craig and Sinclair, "*Kalâm* Cosmological Argument," p. 103.

11. Ravi Zacharias, *The End of Reason: A Response to the New Atheists* (Grand Rapids, MI: Zondervan, 2008), p. 31.

12. Dinesh D'Souza, *What's So Great about Christianity?* (Washington, DC: Regnery, 2007), p. 116.

13. Ibid., p. 126.

14. William Lane Craig, "The Craig-Pigliucci Debate: Does God Exist?" http://www.leaderu.com/offices/billcraig/docs/craig-pigliucci0.html (accessed February 3, 2011).

15. Fred Hoyle, *Astronomy and Cosmology* (San Francisco: W. H. Freeman, 1975), p. 658.

16. Craig, "Craig-Pigliucci Debate."

17. Stephen Hawking and Roger Penrose, "The Singularities of Gravitational Collapse," *Proceedings of the Royal Society of London A* 314 (1970): 529–48.

18. William Lane Craig, "Philosophical and Scientific Pointers to Creatio Ex Nihilo," *Journal of the American Scientific Affiliation* 32, no. 1 (1980): 5–13.

19. Richard Gott III et al., "Will the Universe Expand Forever?" *Scientific American*, March 1976, 65.

20. Stephen Hawking, *A Brief History of Time: From the Big Bang to Black Holes* (New York: Bantam, 1988), p. 50.

21. Ibid.

22. Ibid.

23. D'Souza, *What's So Great about Christianity?*

24. Hawking, *Brief History of Time*, p. 50.

25. William Lane Craig, "Five Arguments for God," *Gospel Coalition*, http://thegospelcoalition.org/publications/cci/five_arguments_for_god/ (accessed May 21, 2010).

26. Arvind Borde, Alan Guth, and Alexander Vilenkin, "Inflationary Spacetimes Are Not Past-Complete," *Physical Review Letters* 90 (2003): 151301.

27. Anthony Aguirre and Steven Gratton Aguirre, "Inflation without a Beginning: A Null Boundary Proposal," *Physical Review D* 67 (2003): 083516.

28. Alexander Vilenkin, e-mail communication with the author, May 20, 2010.

29. Anthony Aguirre, e-mail communications with the author, May 21–22, 2010.

30. Alexander Vilenkin, e-mail communication with the author, May 21, 2010.

31. Sean Carroll and Jennifer Chen, "Spontaneous Inflation and the Origin of the Arrow of Time," Cornell University Library, last modified October 27, 2004, http://arxiv.org/abs/hep-th/0410270 (accessed May 22, 2010).

32. Sean Carroll, *From Eternity to Here: The Quest for the Ultimate Theory of Time* (New York: Dutton, 2010).

33. Sean Carroll, e-mail correspondence with the author, May 21, 2010.

34. For a discussion of the arrow of time, see Victor J. Stenger, *Timeless Reality: Symmetry, Simplicity, and Multiple Universes* (Amherst, NY: Prometheus Books, 2000).

35. Craig and Sinclair, "*Kalâm* Cosmological Argument," p. 157.

36. Anthony Aguirre, e-mail communication with the author, August 20, 2010.

37. Anthony Kenny, *The Five Ways: St. Thomas Aquinas' Proofs of God's Existence* (New York: Schocken Books, 1969), p. 66.

38. Craig, "Craig-Pigliucci Debate."

39. David Atkatz and Heinz Pagels, "Origin of the Universe as Quantum Tunneling Event," *Physical Review D* 25 (1982): 2065–73; James Hartle and Stephen Hawking, "Wave Function of the Universe," *Physical Review D* 28 (1983): 2960–75; Andrei Linde, "Quantum Creation of the Inflationary Universe," *Lettere al Nuovo Cimento* 39

(1984): 401–5; Alexander Vilenkin, "Boundary Conditions and Quantum Cosmology," *Physical Review D* 33 (1986): 3560–69.

40. D'Souza, *What's So Great about Christianity?* p. 116.

41. Dinesh D'Souza, *Life after Death: The Evidence* (Washington, DC: Regenery, 2009), p. 82.

42. Ibid., p. 83.

43. David Atkatz, "Quantum Cosmology for Pedestrians," *American Journal of Physics* 62 (1994): 619–26.

44. Atkatz and Pagels, "Origin of the Universe as a Quantum Tunneling Event."

45. Bryce Dewitt, "Quantum Theory of Gravity: The Canonical Theory," *Physical Review* 160 (1967): 1113–48.

46. Alexander Vilenkin, "Boundary Conditions in Quantum Cosmology," *Physical Review D* 33 (1986): 3560–69.

47. Hartle and Hawking, "Wave Function of the Universe."

48. Stephen Hawking, "Quantum Cosmology," in *Three Hundred Years of Gravitation*, ed. Stephen Hawking and Werner Israel (Cambridge, UK; New York: Cambridge University Press, 1987), 642.

49. Victor J. Stenger, *Timeless Reality: Symmetry, Simplicity, and Multiple Universes* (Amherst, NY: Prometheus Books, 2000), pp. 99–100; Victor J. Stenger, "Time's Arrow Points Both Ways: The View from Nowhen," *Skeptical Inquirer* 8, no. 4 (2001): 90–95; Victor J. Stenger, *The Comprehensible Cosmos: Where Do the Laws of Physics Come From?* (Amherst, NY: Prometheus Books, 2006), pp. 312–19; Victor J. Stenger, "A Scenario for a Natural Origin of Our Universe," *Philo* 9, no. 2 (2006): 93–102.

50. Carroll, *From Eternity to Here*, pp. 353–55.

51. Ibid., p. 354.

CHAPTER 7: GRAVITY IS FICTION

1. Robin Collins, "The Teleological Argument: An Exploration of the Fine-Tuning of the Universe," in *The Blackwell Companion to Natural Theology*, ed. William Lane Craig and James Porter Moreland (Chichester, UK; Malden, MA: Wiley-Blackwell, 2009), 214.

2. The static electrical force given by Coulomb's law is only a

special case of the more general electromagnetic force, which includes magnetic forces as well.

3. For a discussion of gauge invariance and many of the other technical topics in this book at an undergraduate mathematical level, see Victor J. Stenger, *The Comprehensible Cosmos: Where Do the Laws of Physics Come From?* (Amherst, NY: Prometheus Books, 2006).

4. The bodies in our everyday experience are made of particles that have kinetic energy that technically adds to the masses of the bodies. But that is generally negligible compared to the masses of the constituents.

5. Yoichi Iwasaki, "The CP-PACS Project and Lattice QCD Results," *Progress in Theoretical Physics Supplement* 138 (2000): 1–10.

CHAPTER 8: CHEMISTRY

1. John Barrow, "Chemistry and Sensitivity," in *Fitness of the Cosmos for Life*, ed. John Barrow, Simon Conway Morris, Stephen Freeland, and Charles Harper (Cambridge, UK; New York: Cambridge University Press, 2007), 132–49.

CHAPTER 9: THE HOYLE RESONANCE

1. Fred Hoyle et al., "A State in C^{12} Predicted from Astronomical Evidence," *Physical Review Letters* 92 (1953): 1099.

2. Ibid.

3. Fred Hoyle, "The Universe: Past and Present Reflections," *Engineering and Science* (November 1981): 8–12.

4. Helge Kragh, "When Is a Prediction Anthropic? Fred Hoyle and the 7.65 MeV Carbon Resonance," *Philosophy of Science Archive*, http://philsci-archive.pitt.edu/archive/00005332/01/3alphaphil.pdf (accessed May 29, 2010).

5. Steven Weinberg, "Living in the Multiverse," in *Universe or Multiverse?* ed. Bernard Carr (Cambridge, UK: Cambridge University Press, 2007).

6. Mario Livio et al., "The Anthropic Significance of the Existence of an Excited State of C^{12}," *Nature* 340 (1989): 281–89.

7. Weinberg, "Living in the Multiverse."

8. Craig Hogan, "Why the Universe Is Just So," *Reviews of Modern Physics* 72, no. 4 (2000): 1149–61.

CHAPTER 10: PHYSICS PARAMETERS

1. Yoichi Iwasaki, "The CP-PACS Project and Lattice QCD Results," *Progress in Theoretical Physics Supplement* 138 (2000): 10.

2. Robert Jaffe et al., "Quark Masses: An Environmental Impact Statement," *Physical Review D* 79 (2009): 065014.

3. Ibid.

4. Rich Deem, "Evidence for the Fine-Tuning of the Universe," God and Science, http://www.godandscience.org/apologetics/designun.html (accessed November 27, 2008).

5. Since temperature is a measure of the average kinetic energy of the particles in a medium, it can be expressed in energy units.

6. Edward Kolb and Michael Stanley Turner, *The Early Universe* (Reading, MA: Addison-Wesley, 1990), p. 810.

7. Ibid., pp. 70–72.

8. Ibid., p. 90.

9. Roni Harnik et al., "A Universe without Weak Interactions," *Physical Review D* 74 (2006): 035006.

10. Ibid.

11. L. Clavelli and R. E. White III, "Problems in a Weakless Universe," Cornell University Library, submitted September 5, 2006, http://arxiv.org/abs/hep-ph/0609050 (accessed February 8, 2011).

12. Roni Harnik, e-mail communication with the author, August 25, 2010.

13. Stanley Miller, "A Production of Amino Acids under Possible Primitive Earth Conditions," *Science* 117 (1953): 528–29.

14. Martin Rees, *Just Six Numbers: The Deep Forces That Shape the Universe* (New York: Basic Books, 2000), pp. 54–510. First published in 1999.

15. The numbers I use in this section should not be regarded as final since measurements are constantly improving. Regard them as illustrative, which adequately serves my purposes in discussing fine-tuning.

16. D. I. Kazakov, "Beyond the Standard Model (in Search of Supersymmetry)," paper presented at the European School of High Energy Physics, Caramulo, Portugal, August–September 2000.

17. Hugh Ross, "Big Bang Model Refined by Fire," in *Mere Creation: Science, Faith & Intelligent Design*, ed. William Dembski (Downers Grove, IL: InterVarsity Press, 1988), 374.

18. Donald Perkins, *Particle Astrophysics*, 2nd ed. (Oxford; New York: Oxford University Press, 2009), p. 910.

19. Super-Kamiokande Collaboration, "Search for Proton Decay via $p \rightarrow e^{+} \pi^{0}$ in a Large Water Cherenkov Detector," *Physical Review Letters* 81 (1998): 33110. I was involved in this experiment.

20. Thanks to Bob Zannelli for working this out for me.

21. Unless some form of life is impervious to radiation.

22. Ibid.

23. Andrei Sakharov, "Violation of CP Invariance, C Asymmetry, and Baryon Asymmetry of the Universe," *Piz'ma v ZhETF* 5, no. 1 (1967): 32–35.

CHAPTER 11: COSMIC PARAMETERS

1. Hugh Ross, "Big Bang Model Refined by Fire," in *Mere Creation: Science, Faith & Intelligent Design*, ed. William Dembski (Downers Grove, IL: InterVarsity Press, 1988), 373.

2. For nice discussions of this and other cosmological issues, go to Ned Wright's Cosmology Tutorial website at http://www.astro.ucla.edu/~wright/cosmolog.htm (accessed July 21, 2009).

3. Ned Wright, "Big Bang Nucleosynthesis," http://www.astro.ucla.edu/~wright/BBNS.html (accessed April 25, 2010).

4. Ross, "Big Bang Model Refined by Fire," p. 374.

5. William Lane Craig, "The Craig-Pigliucci Debate: Does God Exist?" http://www.leaderu.com/offices/billcraig/docs/craig-pigliucci1.html (accessed February 13, 2010).

6. Dinesh D'Souza, *Life after Death: The Evidence* (Washington, DC: Regenery, 2009), p. 84.

7. Stephen Hawking, *A Brief History of Time: From the Big Bang to Black Holes* (New York: Bantam, 1988), pp. 121–22.

8. Ibid., p. 121.

9. You may wonder how space, which is an empty void, can be "expanding." This is just the language used to describe what happens in the mathematical model. There we find the variable a in the Friedmann equations (see chapter 5), which multiplies the distance between two objects that are "comoving," that is, objects that are at rest with respect to each other. In our universe, this quantity increases with time, given the expansion of the universe.

10. Ross, "Big Bang Model Refined by Fire," p. 373.

11. Peter Schneider, *Extragalactic Astronomy and Cosmology: An Introduction* (Berlin; New York: Springer, 2006), p. 163.

12. Martin Rees, *Just Six Numbers: The Deep Forces That Shape the Universe* (New York: Basic Books, 2000), p. 1111. First published in 1999.

13. Alan Guth, *The Inflationary Universe: The Quest for a New Theory of Cosmic Origins* (Reading, MA: Addison-Wesley Publishing, 1997), p. 242.

14. Rees, *Just Six Numbers*, p. 1211.

15. Paul Davies, *The Goldilocks Enigma: Why Is the Universe Just Right for Life* (London: Allen Lane, 2006), pp. 145–411.

16. Max Tegmark et al., "Towards a Refined Cosmic Concordance Model: Joint 11-Parameter Constraints from the Cosmic Microwave Background and Large-Scale Structure," *Physical Review D* 63 (2000): 043007.

17. E. Komatsu et al., "WMAP Five-Year Observations: Cosmological Interpretation," *Astrophysical Journal Supplement Series* 180 (2009): 330–76.

18. Ibid.

19. Anthony Aguirre, "Cold Big-Bang Cosmology as a Counter Example to Several Anthropic Arguments," *Physical Review D* 64 (2001): 0835011.

20. Anthony Aguirre, "The Cosmic Background Radiation in a Cold Big Bang," *Astrophysical Journal* 533 (2000): 1–111.

CHAPTER 12: THE COSMOLOGICAL CONSTANT

1. S. S. Schweber, *QED and the Men Who Made It: Dyson, Feynman, Schwinger, and Tomonaga* (Princeton, NJ: Princeton University Press, 1994).

2. Gerard 't Hooft, "Dimensional Reduction in Quantum Gravity," Cornell University Library, last modified March 20, 2009, http://lanl.arxiv.org/abs/gr-qc/9310026 (accessed May 30, 2010).

3. For a complete review, see Raphael Bouso, "The Holographic Principle," *Reviews of Modern Physics* 74 (2002): 825–75.

4. D. H. Hsu, "Entropy Bounds and Dark Energy," *Physics Letters B* 594 (2004): 13–16.

5. Andrei Sakharov, "Vacuum Quantum Fluctuations in Curved Space," *Doklady Akademii Nauk SSSR* 177, no. 1 (1967): 70–71.

6. Andrei Linde, "The Inflationary Universe," *Reports on Progress in Physics* 47 (1984): 925–86.

7. See, for example, J. W. Moffat, "Charge Conjugation Invariance of the Vacuum and the Cosmological Constant Problem," *Physics Letters B* 627 (2005): 9–17.

8. Robert Klauber, "Mechanism for Vanishing Zero-Point Energy," Cornell University Library, last modified July 19, 2007, http://lanl.arxiv.org/abs/astro-ph/0309679v3 (accessed August 5, 2010).

9. Richard Feynman, "The Theory of Positrons," *Physical Review* 79 (1949): 749.

10. Alexander Vilenkin, *Many Worlds in One: The Search for Other Universes* (New York: Hill and Wang, 2006). Alejandro Jenkins and Gilad Perez, "Looking for Life in the Multiverse: Universes with Different Physical Laws Might Still Be Habitable," *Scientific American*, January 2010, 42–49.

11. Vilenkin, *Many Worlds in One*, pp. 141–51.

12. Alexander Vilenkin, "Predictions from Quantum Cosmology," *Physical Review Letters* 74 (1995): 846.

13. Vilenkin, *Many Worlds in One*, p. 146.

14. Stephen Feeney et al., "First Observational Test of Eternal Inflation," http://archiv.org/pdf/1012.1995v2 (accessed January 23, 2011).

15. A. G. Riess et al., "Observational Evidence from Supernovae for an Accelerating Universe and a Cosmological Constant," *Astronomical Journal* 116 (1998): 1009; S. Perlmutter et al., "Measurements of Omega and Lambda from 42 High-Redshift Supernovae," *Astrophysical Journal* 517 (1999): 565.

16. Owen Gingrich, "'God's Good,' and the Universe That Knew We Were Coming," in *Science and Religion: Are They Compatible?* ed. Paul Kurtz et al. (Amherst, NY: Prometheus Books, 2003), 51–65.

17. Dinesh D'Souza, *Life after Death: The Evidence* (Washington, DC: Regenery, 2009), pp. 88–89.

18. Don Page, "Preliminary Inconclusive Hint of Evidence against Optimal Fine-Tuning of the Cosmological Constant for Maximizing the Fraction of Baryons Becoming Life," Cornell University Library, last modified January 28, 2011, http://xxx.lanl.gov/abs/1101.2444 (accessed January 20, 2011); Don Page, "Does God So Love the Multiverse," in *Science and Religion in Dialogue, Volume 1*, ed. M. Y. Stewart (Chichester, UK: 2010), 296–410.

CHAPTER 13: MONKEYGOD

1. MonkeyGod can currently be found at http://www.phys.hawaii.edu/vjs/www/monkey.html (accessed February 9, 2011).

2. Victor J. Stenger, *The Unconscious Quantum: Metaphysics in Modern Physics and Cosmology* (Amherst, NY: Prometheus Books, 1995).

3. William Press and Alan Lightman, "Dependence of Macrophysical Phenomena on the Values of the Fundamental Constants," *Philosophical Transactions of the Royal Society A* 310 (1983): 323–36.

4. Bernard Carr and Martin Rees, "The Anthropic Principle and the Structure of the Physical World," *Nature* 278 (1979): 606–13.

5. Stenger, *Unconscious Quantum*, p. 238.

6. Robin Collins, "The Teleological Argument: An Exploration of the Fine-Tuning of the Universe," in *The Blackwell Companion to Natural Theology*, ed. William Lane Craig and James Porter Moreland (Chichester, UK; Malden, MA: Wiley-Blackwell, 2009).

7. Ibid.

8. John Barrow and Frank Tipler, *The Anthropic Cosmological Principle* (Oxford; New York: Oxford University Press, 1986), p. 326.

9. Fred Adams, "Stars in Other Universes: Stellar Structure with Different Fundamental Constants," *Journal of Cosmology and Astroparticle Physics* 2008 (August 2008).

10. Roni Harnik et al., "A Universe without Weak Interactions," *Physical Review D* 74 (2006): 035006.

11. Anthony Aguirre, "Cold Big-Bang Cosmology as a Counter Example to Several Anthropic Arguments," *Physical Review D* 64 (2001): 083508.

CHAPTER 14: PROBABILITY

1. Francis Collins, *The Language of God: A Scientist Presents Evidence for Belief* (New York: Free Press, 2006), p. 74.

2. Ibid., p. 73.

3. Roger Penrose, *The Emperor's New Mind: Concerning Computers, Minds, and the Laws of Physics* (Oxford; New York: Oxford University Press, 1989), p. 343.

4. Wikipedia, "Bayes' Theorem," last modified February 13, 2011, http://en.wikipedia.org/wiki/Bayes'_theorem (accessed February 17, 2011).

5. Stephen Unwin, *The Probability of God: A Simple Calculation That Proves the Ultimate Truth* (New York: Crown Forum, 2003).

6. Larry Ford, e-mail communication with the author, June 2010.

7. This section is adapted from my Reality Check column, "God and Rev. Bayes," *Skeptical Briefs* 17, no. 2 (2007).

8. Michael Ikeda and Bill Jefferys, "The Anthropic Principle Does Not Support Supernaturalism," in *The Improbability of God*, by Michael Martin and Ricki Monnier (Amherst, NY: Prometheus Books, 2006), 150–66. The text is also available on Talk Reason, http://www.talkreason.org/articles/super.cfm (accessed June 3, 2010).

9. Ibid.

10. William Thompson and Edward Schumann, "Interpretation of Statistical Evidence in Criminal Trials: The Prosecutor's Fallacy and the Defense Attorney's Fallacy," *Law and Human Behavior* 2, no. 3 (1987): 167.

11. Bill Jeffreys, e-mail communication with the author, June 12, 2010.

12. Elliott Sober, "The Design Argument," http://philosophy.wisc.edu/sober/design%20argument%2011%202004.pdf (accessed June 13, 2010). An earlier version appears in *The Blackwell Companion to Philosophy of Religion*, edited by W. Mann (London: Blackwell, 2004).

13. Elliott Sober, "Various Topics in Philosophy of Science," http://philosophy.wisc.edu/sober/recent.html (accessed June 13, 2010).

CHAPTER 15: QUANTUM AND CONSCIOUSNESS

1. John Barrow and Frank Tipler, *The Anthropic Cosmological Principle* (Oxford; New York: Oxford University Press, 1986), p. 22.

2. Pew Forum, "Religion among the Millennials: Less Religiously Active Than Older Americans, but Fairly Traditional in Other Ways," Pew Forum on Religion & Public Life, posted February 17, 2010, http://pewforum.org/Age/Religion-Among-the-Millennials.aspx (accessed March 22, 2010).

3. Marilyn Ferguson, *The Aquarian Conspiracy: Personal and Social Transformation in the 1980s* (Los Angeles; New York: J. P. Tarcher, 1980).

4. Robert Basil, ed., *Not Necessarily the New Age: Critical Essays* (Amherst, NY: Prometheus Books, 1988).

5. Fritjof Capra, *The Tao of Physics: An Exploration of the Parallels between Modern Physics and Eastern Mysticism* (London: Wildwood House, 1975).

6. For my review of Capra, see Victor J. Stenger, *Quantum Gods: Creation, Chaos, and the Search for Cosmic Consciousness* (Amherst, NY: Prometheus Books, 2009), pp. 49–54.

7. Amit Goswami, "Physics within Nondual Consciousness," *Philosophy East and West* 51, no. 4 (2001): 535–44.

8. Deepak Chopra, *Life after Death: The Burden of Proof* (New York: Harmony Books, 2006), p. 222.

9. Chris Frith, *Making up the Mind: How the Brain Creates Our Mental World* (Oxford: Blackwell, 2007).

10. Robert Lanza and Deepak Chopra, "Evolution Reigns, but Darwin Outmoded," *San Francisco Chronicle*, October 5, 2009. This is also available through the *Huffington Post*, http://www.huffington post.com/deepak-chopra/evolution-reigns-but-darw_b_309586.html (accessed July 8, 2010).

11. Robert Lanza and Bob Berman, *Biocentrism: How Life and Consciousness Are the Keys to Understanding the True Nature of the Universe* (Dallas, TX: BenBella Books, 2009).

12. Robert Lanza, "The Wise Science," *Humanist* 52, no. 6 (1992): 24.

13. Victor J. Stenger, "The Myth of Quantum Consciousness," *Humanist* 53, no. 3 (1992): 13–15. This is also available at http://www.colorado.edu/philosophy/vstenger/Quantum/Quantum Consciousness.pdf (accessed December 15, 2009).

14. Vinod Wadhawan and Ajita Kamal, "Biocentrism Demystified: A Response to Deepak Chopra and Robert Lanza's Notion of a Conscious Universe," *Nirmukta*, http://nirmukta.com/2009/12/14/biocentrism-demystified-a-response-to-deepak-chopra-and-robert-lanzas-notion-of-a-conscious-universe/#more-2128 (accessed December 15, 2009).

15. Deepak Chopra, *Quantum Healing: Exploring the Frontiers of Mind/Body Medicine* (New York: Bantam Books, 1989); Deepak Chopra, *Ageless Body, Timeless Mind: The Quantum Alternative to Growing Old* (New York: Harmony Books, 1993).

16. William Arntz et al., *What the Bleep Do We Know!? Discovering the Endless Possibilities for Altering Your Everyday Reality* (Deerfield Beach, FL: Health Communications, 2005).

17. Rhonda Byrne, *The Secret* (New York: Attria Books, 2006).

18. Ibid., p. 4.

19. Ibid., pp. 9–10.

20. Ibid., p. 15.

21. Ibid.

22. See "Iconic Photos, Solvay Conference," posted January 28, 2010, http://iconicphotos.wordpress.com/2010/01/28/the-solvay-conference/ (accessed July 7, 2010).

23. Victor J. Stenger, *The Unconscious Quantum: Metaphysics in Modern Physics and Cosmology* (Amherst, NY: Prometheus Books, 1995), pp. 66–79.

24. Albert Einstein et al., *The Born-Einstein Letters: Correspondence between Albert Einstein and Max and Hedwig Born from 1916–1955, with Commentaries by Max Born* (London: Macmillan, 1971).

25. Michael Polanyi, *The Logic of Personal Knowledge: Essays Presented to Michael Polanyi on His Seventieth Birthday, 11th March 1961* (London: Routledge & Paul, 1961).

26. Richard Elvee and John Archibald Wheeler, *Mind in Nature* (San Francisco: Harper & Row, 1982), p. 17.

27. Paul Davies and John Gribbin, *The Matter Myth: Dramatic Discoveries That Challenge Our Understanding of Physical Reality* (New York: Simon & Schuster, 2007), p. 215.

CHAPTER 16: SUMMARY AND REVIEW

1. Rich Deem, "Evidence for the Fine-Tuning of the Universe," God and Science, http://www.godandscience.org/apologetics/designun.html (accessed November 27, 2008).

2. Hugh Ross, "Big Bang Model Refined by Fire," in *Mere Creation: Science, Faith & Intelligent Design*, ed. William Dembski (Downers Grove, IL: InterVarsity Press, 1988), 374.

3. Stephen Hawking, *A Brief History of Time: From the Big Bang to Black Holes* (New York: Bantam, 1988), pp. 121–22.

4. Ibid., p. 121.

5. Lee Smolin, "Scientific Alternatives to the Anthropic Principle," Cornell University Library, last modified July 29, 2004, http://arxiv.org/abs/hep-th/0407213v3 (accessed February 11, 2011).

6. Stephen Unwin, *The Probability of God: A Simple Calculation That Proves the Ultimate Truth* (New York: Crown Forum, 2003).

7. Michael Ikeda and Bill Jefferys, "The Anthropic Principle Does Not Support Supernaturalism," in *The Improbability of God*, by Michael Martin and Ricki Monnier (Amherst, NY: Prometheus Books, 2006), 150–66. The text is also available on Talk Reason, http://www.talkreason.org/articles/super.cfm (accessed June 3, 2010).

8. Richard Swinburne, "Argument from the Fine-Tuning of the Universe," in *Modern Cosmology & Philosophy*, ed. John Leslie (Amherst, NY: Prometheus Books, 1999), 154–73.

9. Don Page, "Preliminary Inconclusive Hint of Evidence against Optimal Fine-Tuning of the Cosmological Constant for Maximizing the Fraction of Baryons Becoming Life," Cornell University Library, last modified January 28, 2011, http://xxx.lanl.gov/abs/1101.2444 (accessed February 11, 2011).

Bibliography

Adams, Fred. "Stars in Other Universes: Stellar Structure with Different Fundamental Constants." *Journal of Cosmology and Astroparticle Physics* (August 2008).

Aguirre, Anthony. "Cold Big-Bang Cosmology as a Counter Example to Several Anthropic Arguments." *Physical Review D* 64 (2001): 083508.

———. "The Cosmic Background Radiation in a Cold Big Bang." *Astrophysical Journal* 533 (2000): 1–18.

Aguirre, Anthony, and Steven Gratton. "Inflation without a Beginning: A Null Boundary Proposal." *Physical Review D* 67 (2003): 083515.

Atkatz, David. "Quantum Cosmology for Pedestrians." *American Journal of Physics* 62, no. 7 (1994): 619–27.

Atkatz, David, and Heinz Pagels. "Origin of the Universe as a Quantum Tunneling Event." *Physical Review Letters D* 25 (1982): 2065–73.

Barrow, John. "Chemistry and Sensitivity." In *Fitness of the Cosmos for Life: Biochemistry and Fine-Tuning*, edited by John Barrow, Simon Conway Morris, Stephen Freeland, and Charles Harper, 132–50. Cambridge, UK: Cambridge University Press, 2008.

———. *The Constants of Nature: From Alpha to Omega*. London: Jonathan Cape, 2002.

———. "Cosmology and the Origin of Life." Paper presented at the Conference on the Origin of Intelligent Life in the Universe, Varenna, Italy, September 30, 1998.

Barrow, John, and Frank Tipler. *The Anthropic Cosmological Principle*. Oxford; New York: Oxford University Press, 1986.

Barrow, John, Simon Conway Morris, Stephen Freeland, and Charles Harper, eds. *Fitness of the Cosmos for Life: Biochemistry and Fine-Tuning*. Cambridge, UK; New York: Cambridge University Press, 2008.

Bartholomew, David. *God, Chance, and Purpose: Can God Have It Both Ways?* Cambridge, UK; New York: Cambridge University Press, 2008.

Basil, Robert, ed. *Not Necessarily the New Age: Critical Essays*. Amherst, NY: Prometheus Books, 1988.

Begley, Sharon. "Science Finds God." *Newsweek*, July 20, 1998, 46.

Bekenstein, Jacob. "The Second Law of Thermodynamics." *Letters to Nuovo Cimento* 4, no. 15 (1972): 737–40.

Bertola, F., and Umberto Curi. *The Anthropic Principle: Proceedings of the Second Venice Conference on Cosmology and Philosophy*. Cambridge, UK: Cambridge University Press, 1988.

Birx, H. James, ed. *Encyclopedia of Time: Science, Philosophy, Theology, & Culture*. Los Angeles; London: Sage Publications, 2009.

Bohm, David, and B. J. Hiley. *The Undivided Universe: An Ontological Interpretation of Quantum Theory*. London; New York: Routledge, 1993.

Borde, Arvind, Alan Guth, and Alexander Vilenkin. "Inflationary Spacetimes Are Not Past-Complete." *Physical Review Letters* 90 (2003): 151301.

Bostrom, Nick. *Anthropic Bias: Observation Selection Effects in Science and Philosophy*. New York: Routledge, 2002.

Bouso, Raphael. "The Holographic Principle." *Reviews of Modern Physics* 74 (2002): 825–75.

Breuer, Reinhard. *The Anthropic Principle: Man as the Focal Point of Nature*. Boston: Birkhäuser, 1991.

Byers, Nina. "Emmy Noether." In *Out of the Shadows: Contributions of Twentieth-Century Women to Physics*, edited by Nina Byers and Gary Williams. Cambridge, UK; New York: Cambridge University Press, 2006.

Carr, Bernard. "On the Origin, Evolution, and the Purpose of the Physical Universe." In *Modern Cosmology and Philosophy*, edited by John Leslie, 140–59. Amherst, NY: Prometheus Books, 1998.

———, ed. *Universe or Multiverse*. Cambridge, UK: Cambridge University Press, 2007.

Carr, Bernard, and Martin Rees. "The Anthropic Principle and the Structure of the Physical World." *Nature* 278 (1979): 606–12.

Carroll, Sean. *From Eternity to Here: The Quest for the Ultimate Theory of Time*. New York: Dutton, 2010.

Carroll, Sean, and Jennifer Chen. "Spontaneous Inflation and the Origin of the Arrow of Time." Cornell University Library. Last modified October 27, 2004. http://arxiv.org/abs/hep-th/0410270.

Carter, Brandon. "Large Number Coincidences and the Anthropic Principle in Cosmology." In *Confrontation of Cosmological Theory with Astronomical Data*, edited by M. S. Longair, 291–98. Derdrecht: Reidel, 1974.

Chopra, Deepak. *Ageless Body, Timeless Mind: The Quantum Alternative to Growing Old*. New York: Harmony Books, 1993.

———. *Quantum Healing: Exploring the Frontiers of Mind/Body Medicine*. New York: Bantam Books, 1989.

Circkovic, Milan, and Nick Bostrom. "Cosmological Constant and the Final Anthropic Hypothesis." *Astrophysics and Space Science* 274 (2000): 675–87.

Clavelli, L., and R. E. White III. "Problems in a Weakless Universe." Cornell University Library. Submitted September 5, 2006. http://arxiv.org/abs/hep-ph/0609050.

Cohen, Andrew, David Kaplan, and Ann Nelson. "Effective Field Theory, Black Holes, and the Cosmological Constant." *Physical Review Letters* 82 (1999): 4971–74.

Cole, David. "The Chinese Room Argument." Stanford Encyclopedia of Philosophy. Last modified September 22, 2009. http://plato.stanford.edu/entries/chinese-room/#6.

Collins, Robin. "The Teleological Argument: An Exploration of the Fine-Tuning of the Universe." In *The Blackwell Companion to Natural Theology*, edited by William Lane Craig and James Porter Moreland, 202–81. Chichester, UK; Malden, MA: Wiley-Blackwell, 2009.

Comings, David. *Did Man Create God? Is Your Spiritual Brain at Peace with Your Thinking Brain?* Duarte, CA: Hope Press, 2008.

Corey, Michael Anthony. *God and the New Cosmology: The Anthropic Design Argument*. Lanham, MD: Rowman & Littlefield, 1993.

———. *The God Hypothesis : Discovering Design in Our "Just Right" Goldilocks Universe*. Lanham, MD: Rowman & Littlefield, 2001.

Craig, William Lane, "Five Arguments for God." Gospel Coalition. http://thegospelcoalition.org/publications/cci/five_arguments_for_god.

———. *The Kalâm Cosmological Argument. Library of Philosophy and Religion*. London: Macmillan, 1979.

Craig, William Lane, and James Porter Moreland, eds. *The Blackwell Companion to Natural Theology*. Chichester, UK; Malden, MA: Wiley-Blackwell, 2009.

Craig, William Lane, and James Sinclair. "The *Kalâm* Cosmological Argument." In *The Blackwell Companion to Natural Theology*, edited by William Lane Craig and James Porter Moreland, 101–201. Chichester, UK; Malden, MA: Wiley-Blackwell, 2009.

Darwin, Charles. *On the Origin of Species: By Means of Natural Selection, or the Preservation of Favoured Races in the Struggle for Life*. Edited by David Quammen. New York: Sterling Publishing, 2008.

Davies, Paul. *The Cosmic Blueprint: New Discoveries in Nature's Creative Ability to Order the Universe*. Philadelphia, PA: Templeton Foundation Press, 2004.

———.*Cosmic Jackpot: Why Our Universe Is Just Right for Life*. Boston: Houghton Mifflin, 2007.

———. *The Goldilocks Enigma: Why Is the Universe Just Right for Life?* London: Allen Lane, 2006.

Dawkins, Richard. *The Greatest Show on Earth: The Evidence for Evolution*. New York: Free Press, 2009.

de Bernardis, P., et al. "A Flat Universe from High Resolution Maps of the Cosmic Microwave Background." *Nature* 404 (2000): 955–59.

Deem, Rich. "Evidence for the Fine-Tuning of the Universe." Evidence for God. http://www.godandscience.org/apologetics/designun.html.

Dembski, William. *Intelligent Design: The Bridge between Science & Theology*. Downers Grove, IL: InterVarsity Press, 1999.

———, ed. *Mere Creation: Science, Faith & Intelligent Design*. Downers Grove, IL: InterVarsity Press, 1998.

Dennett, Daniel, and Alvin Plantinga. *Science and Religion: Are They Compatible?* New York: Oxford University Press, 2011.

Dewitt, Bryce. "Quantum Theory of Gravity: The Canonical Theory." *Physical Review* 160 (1967): 1113–48.

Dicke, Robert. "Dirac's Cosmology and Mach's Principle." *Nature* 192 (1961): 440.

Dirac, Paul. "The Cosmological Constants." *Nature* 139 (1937): 323.

———. *The Principles of Quantum Mechanics*. Oxford: Oxford University Press, 1989.

D'Souza, Dinesh. *Life after Death: The Evidence*. Washington, DC: Regenery, 2009.

———. *What's So Great about Christianity?* Washington, DC: Regenery, 2007.

Dyson, Freeman. "Feynman's Proof of Maxwell's Equations." *American Journal of Physics* 58, no. 3 (1989): 543–44.

Easther, Richard, and David Lowe. "Holography, Cosmology, and the Second Law of Thermodynamics." *Physical Review Letters* 82, no. 25 (1999): 4967–70.

Eddington, Arthur. *The Mathematical Theory of Relativity*. London: Cambridge University Press, 1923.

Einstein, Albert, Max Born, and Hedwig Born. *The Born-Einstein Letters: Correspondence between Albert Einstein and Max and Hedwig Born from 1916–1955, with Commentaries by Max Born*. London: Macmillan, 1971.

Elvee, Richard, and John Archibald Wheeler. *Mind in Nature*. San Francisco: Harper & Row, 1982.

Feeney, Stephen, et al. "First Observational Test of Eternal Inflation." http://archiv.org/pdf/1012.1995v2.

Ferguson, Marilyn. *The Aquarian Conspiracy: Personal and Social Transformation in the 1980s*. Los Angeles; New York: J. P. Tarcher, 1980.

Feyman, Richard. *QED: The Strange Theory of Light and Matter*. Princeton, NJ: Princeton University Press, 1985.

———. "The Theory of Positrons." *Physical Review* 79 (1949): 749.

Field, J. H. "Derivation of the Lorentz Force Law and the Magnetic Field Concept Using an Invariant Formulation of the Lorentz

Transformation." Cornell University Library. Submitted July 28, 2003. http://arxiv.org/abs/physics/0307133.

Frith, Chris. *Making up the Mind: How the Brain Creates Our Mental World*. Oxford: Blackwell, 2007.

Greenstein, George. *The Symbiotic Universe: Life and Mind in the Cosmos*. New York: Morrow, 1988.

Gribbin, John, and Martin Rees. *Cosmic Coincidences: Dark Matter, Mankind, and Anthropic Cosmology*. New York: Bantam Books, 1989.

Guth, Alan. "Inflationary Universe: A Possible Solution to the Horizon and Flatness Problems." *Physical Review D* 23, no. 2 (1981): 347–56.

———. *The Inflationary Universe: The Quest for a New Theory of Cosmic Origins*. Reading, MA: Addison-Wesley Publishing, 1997.

Haack, Susan. *Defending Science—Within Reason: Between Scientism and Cynicism*. Amherst, NY: Prometheus Books, 2003.

Hameroff, Stuart. "Quantum Consciousness." http://www.quantum consciousness.org/.

Harnik, Roni, Graham Kribs, and Gilad Perez. "A Universe without Weak Interactions." *Physical Review D* 74 (2006): 035006.

Harris, Errol. *Cosmos and Anthropos: A Philosophical Interpretation of the Anthropic Cosmological Principle*. Atlantic Highlands, NJ: Humanities Press International, 1991.

———. *Cosmos and Theos: Ethical and Theological Implications of the Anthropic Cosmological Principle*. Atlantic Highlands, NJ: Humanities Press, 1992.

Harrison, Edward Robert. *Masks of the Universe*. New York; London: Macmillan; Collier Macmillan, 1985.

Hartle, James, and Stephen Hawking. "Wave Function of the Universe." *Physical Review D* 28 (1983): 2960–75.

Hawking, Stephen. *A Brief History of Time: From the Big Bang to Black Holes*. New York: Bantam Books, 1988.

Hawking, Stephen, and W. Israel, eds. *Three Hundred Years of Gravitation*. Cambridge, UK; New York: Cambridge University Press, 1987.

Hogan, Craig. "Nuclear Astrophysics of Worlds in the String Landscape." *Physical Review D* 74 (2006): 123514.

———. "Why the Universe Is Just So." *Reviews of Modern Physics* 72, no. 4 (2000): 1149–61.

Hoyle, Fred. "The Universe: Past and Present Reflections." *Engineering and Science*, November 1981, 8–12.

Hoyle, Fred, et al. "A State in C^{12} Predicted from Astronomical Evidence." *Physical Review Letters* 92 (1953): 1095.

Hsu, D. H. "Entropy Bounds and Dark Energy." *Physics Letters B* 594 (2004): 13–16.

Hume, David. *An Enquiry concerning Human Understanding*. Edited by P. F. Millican. Oxford; New York: Oxford University Press, 2007.

Ikeda, Michael, and Bill Jeffreys. "The Anthropic Principle Does Not Support Supernaturalism." Talk Reason. Posted January 12, 2004. http://www.talkreason.org/articles/super.cfm.

Iwasaki, Yoichi. "The CP-PACS Project and Lattice QCD Results." *Progress in Theoretical Physics Supplement* 138 (2000): 1–10.

Jackson, Roy. *The God of Philosophy: An Introduction to the Philosophy of Religion*. London: Philosophers' Magazine, 2001.

Jaffe, Robert, Alejandro Jenkins, and Kimchi Itamar. "Quark Masses: An Environmental Impact Statement." *Physical Review D* 79 (2009): 065014.

Jastrow, Robert. *God and the Astronomers*. 2nd ed. New York: Norton, 1992.

Jeffrey, Grant. *Creation: Remarkable Evidence of God's Design*. Toronto: Frontier Research Publications, 2003.

Jones, John, III. "*Kitzmiller v. Dover Area School District, et al.*, Case No. 04cv2688." Filed December 20, 2005. http://www.pamd.uscourts.gov/kitzmiller/kitzmiller_342.pdf.

Kaplan, David, and Raman Sundrum. "A Symmetry for the Cosmological Constant." *Journal of High Energy Physics* 07 (2006): 42.

Kapogiannis, D., et al. "Cognitude and Neural Foundations of Religious Belief." *Proceedings of the National Academy of Sciences USA* 106, no. 12 (2009): 4876–81.

Kazakov, D. I. "Beyond the Standard Model (in Search of Supersymmetry)." Paper presented at the European School of High Energy Physics, Caramulo, Portugal, August–September 2000.

Kazanas, Demos. "Dynamics of the Universe and Spontaneous Symmetry Breaking." *Astrophysical Journal* 241 (1980): L59–L63.

Klapwijk, Jacob. *Purpose in the Living World: Creation and Emergent Evolution*. Cambridge, UK; New York: Cambridge University Press, 2008.

Klauber, Robert. "Mechanism for Vanishing Zero-Point Energy." Cornell University Library. Last modified July 19, 2007. http://lanl.arxiv.org/abs/astro-ph/0309679v3.

Kolb, Edward, and Michael Stanley Turner. *The Early Universe*. Reading, MA: Addison-Wesley, 1990.

Komatsu, E., et al. "Wmap Five-Year Observations: Cosmological Interpretation." *Astrophysical Journal Supplement Series* 180 (2009): 330–76.

Kragh, Helge. "When Is a Prediction Anthropic? Fred Hoyle and the 7.65 MeV Carbon Resonance." http://philsci-archive.pitt.edu/archive/00005332/01/3alphaphil.pdf.

Kuhn, Thomas. *The Structure of Scientific Revolutions*. Chicago: University of Chicago Press, 1962.

Lanza, Robert, and Deepak Chopra. "Evolution Reigns, but Darwin Outmoded." *San Francisco Chronicle*, October 5, 2009.

Leslie, John, ed. *Modern Physics and Cosmology*. Amherst, NY: Prometheus Books, 1998. See esp. "The Anthropic Principle Today," 289–310.

Linde, Andrei. "The Inflationary Universe." *Reports on Progress in Physics* 47 (1984): 925–86.

———. "A New Inflationary Universe Scenario: A Possible Solution of the Horizon, Flatness, Homogeneity, Isotropy, and Primordial Monopole Problems." *Physics Letters B* 108 (1982): 389.

———. "Quantum Creation of the Inflationary Universe." *Physics Letters B* 108 (1982): 389–92.

Livio, M., et al. "The Anthropic Significance of the Existence of an Excited State of C^{12}." *Nature* 340 (1989): 281–84.

Mann, W., ed. *The Blackwell Companion to Philosophy of Religion*. London: Blackwell, 2004.

Martin, Michael, and Ricki Monnier. *The Impossibility of God*. Amherst, NY: Prometheus Books, 2003.

———. *The Improbability of God*. Amherst, NY: Prometheus Books, 2006.

Miller, Stanley. "A Production of Amino Acids under Possible Primitive Earth Conditions." *Science* 117 (1953): 528–29.

Moffat, J. W. "Charge Conjugation Invariance of the Vacuum and the Cosmological Constant Problem." *Physics Letters B* 627 (2005): 9–17.

National Center for Science Education. "Polling Evolution in Three Countries." Posted July 16, 2010. http://ncse.com/news/2010/07/polling-evolution-three-countries-005708/.

Newport, Frank. "Republicans, Democrats Differ on Creationism." Gallup. Posted June 20, 2008. http://www.gallup.com/poll/108226/Republicans-Democrats-Differ-Creationism.aspx.

Ourisson, Guy. "Course-Tuning in the Origin of Life." In *Fitness of the Cosmos for Life: Biochemistry and Fine-Tuning*, edited by John Barrow, Simon Conway Morris, Stephen Freeland, and Charles Harper, 421–39. Cambridge, UK: Cambridge University Press, 2008.

Overman, Dean. *A Case against Accident and Self-Organization*. New York: Rowman & Littlefield, 1997.

Page, Don. "Does God So Love the Multiverse?" In *Science and Religion in Dialogue*, vol. 1, edited by M. Y. Stewart, 296–410. Chichester, UK: 2010.

———. "Preliminary Inconclusive Hint of Evidence against Optimal Fine-Tuning of the Cosmological Constant for Maximizing the Fraction of Baryons Becoming Life." Cornell University Library. Last modified January 28, 2011. http://xxx.lanl.gov/abs/1101.2444.

Paley, William. *Natural Theology; Or, Evidences of the Existence and Attributes of the Deity*. London: Wilks and Taylor, 1802.

Perkins, Donald. *Particle Astrophysics*. 2nd ed. Oxford; New York: Oxford University Press, 2009.

Perlmutter, S., et al. "Measurements of Omega and Lambda from 42 High-Redshift Supernovae." *Astrophysical Journal* 517 (1999): 565.

Pew Forum. "Religion among the Millennials: Less Religiously Active Than Older Americans, but Fairly Traditional in Other Ways." Posted February 17, 2010. http://pewforum.org/Age/Religion-Among-the-Millennials.aspx.

Polanyi, Michael. *The Logic of Personal Knowledge: Essays Presented to Michael Polanyi on His Seventieth Birthday, 11th March 1961*. London: Routledge & Paul, 1961.

Press, William, and Alan Lightman. "Dependence of Macrophysical Phenomena on the Values of the Fundamental Constants." *Philosophical Transactions of the Royal Society of London A* 310 (1983): 323–36.

Preston, John, and Mark Bishop. *Views into the Chinese Room: New Essays on Searle and Artificial Intelligence*. Oxford; New York: Clarendon Press, 2002.

Rees, Martin. *Just Six Numbers: The Deep Forces That Shape the Universe*. New York: Basic Books, 2000.

Riess, A. G., et al. "Observational Evidence from Supernovae for an Accelerating Universe and a Cosmological Constant." *Astronomical Journal* 116 (1998): 1009.

Ross, Hugh. "Big Bang Model Refined by Fire." In *Mere Creation: Science, Faith & Intelligent Design*, edited by William Dembski, 363–83. Downers Grove, IL: InterVarsity Press, 1988.

———. *The Creator and the Cosmos: How the Greatest Scientific Discoveries of the Century Reveal God*. 2nd exp. ed. Colorado Springs, CO: NavPress, 1995.

Rothman, Tony. "A 'What You See Is What You Beget' Theory." *Discover*, May 1987.

Sakharov, Andrei. "Vacuum Quantum Fluctuations in Curved Space." *Doklady Akademii Nauk SSSR* 177, no. 1 (1967): 70–71.

Salpeter, Edwin. "Accretion of Interstellar Matter by Massive Objects." *Astrophysical Journal* 140 (1964): 796–800.

Shannon, Claude, and Warren Weaver. *The Mathematical Theory of Communication*. Urbana: University of Illinois Press, 1998.

Smolin, Lee. "Scientific Alternatives to the Anthropic Principle." Cornell University Library. Last modified July 29, 2004. http://arxiv.org/abs/hep-th/0407213v3.

———. *The Trouble with Physics: The Rise of String Theory, the Fall of a Science, and What Comes Next*. Boston: Houghton Mifflin, 2006.

Smoot, G., and K. Davidson. *Wrinkles in Time*. New York: Morrow, 1993.

Smyth, Piazzi. *The Great Pyramid: Its Secrets and Mysteries Revealed*. New York: Bell Publishing, 1978.

Sober, Elliott. "The Design Argument." http://philosophy.wisc.edu/sober/design%20argument%2011%202004.pdf.

Spergel, D. N., et al. "Wilkinson Microwave Anisotropy Probe (WMAP) Three-Year Results: Implications for Cosmology." *Astrophysical Journal Supplement* 170 (2006): 377.

Steibing, William. *Ancient Astronauts, Cosmic Collisions and Other Popular Theories about Man's Past*. Amherst, NY: Prometheus Books, 1984.

Stenger, Victor J. "The Anthropic Coincidences: A Natural Explanation." *Skeptical Intelligencer* 3, no. 3 (1999): 2–17.

———. *The Comprehensible Cosmos: Where Do the Laws of Physics Come From?* Amherst, NY: Prometheus Books, 2006.

———. *Has Science Found God? The Latest Results in the Search for Purpose in the Universe*. Amherst, NY: Prometheus Books, 2003.

———. "Neutrino Oscillations in DUMAND." Paper presented at the Neutrino Mass Mini-Workshop, Telemark, Wisconsin, 1980.

———. *Quantum Gods: Creation, Chaos, and the Search for Cosmic Consciousness*. Amherst, NY: Prometheus Books, 2009.

———. "A Scenario for the Natural Origin of the Universe." *Philo* 9, no. 2 (2006): 93–102.

———. "Time, Operational Definition of." In *Encyclopedia of Time*, edited by James Birx, 1293–95. Los Angeles; London: Sage Publications, 2009.

———. *Timeless Reality: Symmetry, Simplicity, and Multiple Universes*. Amherst, NY: Prometheus Books, 2000.

———. *The Unconscious Quantum: Metaphysics in Modern Physics and Cosmology*. Amherst, NY: Prometheus Books, 1995.

Super-Kamiokande Collaboration. "Search for Proton Decay Via $p \rightarrow e^{+}\pi^{0}$ in a Large Water Cherenkov Detector." *Physical Review Letters* 81 (1998): 3319.

Susskind, Leonard. *Cosmic Landscape: String Theory and the Illusion of Intelligent Design*. Boston: Little, Brown, 2005.

———. "The World as a Hologram." *Journal of Mathematical Physics* 36 (1995): 6377.

Swinburne, Richard. "Argument from the Fine-Tuning of the Universe." In *Modern Cosmology & Philosophy*, edited by John Leslie, 160–79. Amherst, NY: Prometheus Books, 1998.

Tegmark, Max, and Martin Rees. "Why Is the CMB Fluctuation Level 10{-5}." *Astrophysical Journal* 499 (1998): 526–32.

Tegmark, Max, Matias Zaldarriaga, and Andrew Hamilton. "Towards a Refined Cosmic Concordance Model: Joint 11-Parameter Constraints from the Cosmic Microwave Background and Large-Scale Structure." *Physical Review D* 63 (2000): 043007.

Teilhard de Chardin, Pierre. *The Phenomenon of Man*. New York: Harper, 1959.

Thompson, William, and Edward Schumann. "Interpretation of Statistical Evidence in Criminal Trials: The Prosecutor's Fallacy and the Defense Attorney's Fallacy." *Law and Human Behavior* 2, no. 3 (1987): 167.

't Hooft, Gerard. "Dimensional Reduction in Quantum Gravity." Cornell University Library. Last modified March 20, 2009. http://lanl.arxiv.org/abs/gr-qc/9310026.

Tipler, Frank. *The Physics of Immortality: Modern Cosmology, God, and the Resurrection of the Dead*. New York: Doubleday, 1994.

UCSD, Physics 130. "The Lorentz Force from the Classical Hamiltonian." Posted December 22, 2008. http://quantummechanics.ucsd.edu/ph130a/130_notes/node302.html.

Unwin, Stephen. *The Probability of God: A Simple Calculation That Proves the Ultimate Truth*. New York: Crown Forum, 2003.

Vilenkin, Alexander. "Boundary Conditions in Quantum Cosmology." *Physical Review D* 33 (1986): 3560–69.

———. *Many Worlds in One: The Search for Other Universes*. New York: Hill & Wang, 2006.

———. "Predictions from Quantum Cosmology." *Physical Review Letters* 74 (1995): 846.

Vision Critical. "Americans Are Creationists; Britons and Canadians Side with Evolution." Posted July 15, 2010. http://www.visioncritical.com/2010/07/americans-are-creationists-britons-and-canadians-side-with-evolution/.

Weinberg, Steven. *Gravitation and Cosmology*. New York: Wiley, 1972.

———. "Living in the Multiverse." In *Universe or Multiverse*, edited by Bernard Carr. Cambridge, UK: Cambridge University Press, 2007.

Weyl, Hermann. *Annals of Physics* 59 (1919): 101.

Wikipedia. "Bayes Theorem." Last modified February 13, 2011. http://en.wikipedia.org/wiki/Bayes'_theorem.

Wright, Ned. "Big Bang Nucleosynthesis." Last modified September 12, 2004. http://www.astro.ucla.edu/~wright/BBNS.html.

Zehavi, I., and A. Dekel. "Evidence for a Positive Cosmological Constant from Flows of Galaxies and Distant Supernovae." *Nature* 401 (1999): 252–54.

Index

About the Author

VICTOR J. STENGER grew up in a Catholic working-class neighborhood in Bayonne, New Jersey. His father was a Lithuanian immigrant, his mother the daughter of Hungarian immigrants. He attended public schools and received a bachelor of science degree in electrical engineering from Newark College of Engineering (now New Jersey Institute of Technology) in 1956. While at NCE, he was editor of the student newspaper and received several journalism awards.

Moving to Los Angeles on a Hughes Aircraft Company fellowship, Dr. Stenger received a master of science degree in physics from UCLA in 1959 and a doctorate in physics in 1963. He then took a position on the faculty of the University of Hawaii and retired to Colorado in 2000. He currently is adjunct professor of philosophy at the University of Colorado and emeritus professor of physics at the University of Hawaii. Dr. Stenger is a fellow of the Committee for Skeptical Inquiry and a research fellow of the Center for Inquiry. Dr. Stenger has also held visiting positions on the faculties of the University of Heidelberg in Germany and the University of Oxford in England,

and he has been a visiting researcher at Rutherford Laboratory in England, the National Nuclear Physics Laboratory in Frascati, Italy, and the University of Florence in Italy.

His research career spanned the period of great progress in elementary particle physics that ultimately led to the current *standard model*. He participated in experiments that helped establish the properties of strange particles, quarks, gluons, and neutrinos. He also helped pioneer the emerging fields of very high-energy gamma ray and neutrino astronomy. In his last project before retiring, Dr. Stenger collaborated on the underground experiment in Japan that in 1998 showed for the first time that the neutrino has mass. The Japanese leader of this project, Masatoshi Koshiba, shared the 2002 Nobel Prize for this work.

Victor J. Stenger has had a parallel career as an author of critically well-received popular-level books that interface between physics and cosmology and philosophy, religion, and pseudoscience. His 2007 book, *God: The Failed Hypothesis*, made the New York Times bestseller list in March of that year.

Dr. Stenger and his wife, Phylliss, have been happily married since 1962 and have two children and four grandchildren. They will celebrate their golden wedding anniversary on October 6, 2012. They now live in Lafayette, Colorado, and travel the world as often as they can.

Dr. Stenger maintains a popular website where much of his writing can be found, at http://www.colorado.edu/philosophy/vstenger/. He also maintains an e-mail discussion list, avoid-L, where the topics range from his own writings to the whole gamut of intellectual discourse and politics.

Other Books by Victor J. Stenger

Not by Design: The Origin of the Universe (1988)
Physics and Psychics: The Search for a World beyond the Senses (1990)
The Unconscious Quantum: Metaphysics in Modern Physics and Cosmology (1995)
Timeless Reality: Symmetry, Simplicity, and Multiple Universes (2000)
Has Science Found God? The Latest Results in the Search for Purpose in the Universe (2003)
The Comprehensible Cosmos: Where Do the Laws of Physics Come From? (2006)
God: The Failed Hypothesis—How Science Shows That God Does Not Exist (2007)
Quantum Gods: Creation, Chaos, and the Search for Cosmic Consciousness (2009)
The New Atheism: Taking a Stand for Science and Reason (2009)